Architectures Beyond Boxes and Boundaries

My Life by Design

Larry Bell

Architectures Beyond Boxes and Boundaries: My Life by Design

Also by Larry Bell

Scared Witless: Prophets and Profits of Climate Doom
Cosmic Musings: Contemplating Life beyond Self
Climate of Corruption: Politics and Power behind the Global
Warming Hoax
Reflections on Oceans and Puddles
Reinventing Ourselves: How Technology Is Rapidly and
Radically Transforming Humanity
What Makes Humans Truly Exceptional: Including Us
Cyberwarfare: Targeting America
Thinking Whole: Rejecting Half-Witted Left & Right Brain
Limitations
The Weaponization of AI and the Internet
Beyond Flagpoles and Footprints: Pioneering the Space Frontier

Cover illustration by Albert Rajkumar
Cover Design by Guy D. Corp www.GrafixCorp.com

STAIRWAY≣PRESS

www.StairwayPress.com
1000 West Apache Trail
Suite 126
Apache Junction, AZ 85120

Dedication

Dedicated to those who have inspired and discouraged me; trusted and doubted me; joined and competed with me; helped and obstructed me; welcomed and rejected me; agreed and argued with me; and honored and ridiculed me.

I thank and blame you all for enriching my wonderfully challenging life journey.

Introduction

THE GIRDING PREMISE of this book views life—everyone's—as our most important design activity. Having said this, while I truly do give real credence to the general idea, I also recognize that it is overly simplistic to visualize the process as a singular "project" guided and pursued in accordance with a static master plan.

Intervening events, opportunities, challenges, and setbacks make life design a constant work-in-progress to fully become the person we wish to be. Each change, whether intentional or through serendipity—or for better or worse—warrants and instructs renewed assessments. We become more or less of that ideal person through practice and habit.

Whether or not that person we are constantly in the process of becoming is the same individual we most admire entails some important questions. For example, we might ask ourselves whether we truly care enough about that goal—each goal—to take on all the burdens of responsibility it demands. Are we confident enough about our preparation and abilities to accept worthwhile risks? Have we demonstrated the innovative persistence to prevail over daunting obstacles and to rebound from setbacks?

Of course, neither we nor the circumstances that surround us need to be perfect. If they were, we wouldn't enjoy the wonderful challenges and rich satisfactions of improving them.

As for dealing with those obstacles and setbacks, I am inspired by examples of colleagues and other friends who have endured, survived and achieved far more than I have. Some of their wonderful stories are highlighted in my previous book, *Thinking Whole: In Pursuit of More Mindful Left and Right Brain Lives.*

Life, whether by conscious design or default, tests our overall tolerance of risk in pursuit of purposes, ideas, opportunities, solutions and outcomes we believe in. This includes necessary confidence and conviction to "think out of the box" in offering proposals, ideas and assessments that break with accepted group-think conventions and risk rueful penalties of challenging "expert" opinions.

Just as there are lots of life experiences and design projects that I'm quite proud of, there are also many others that didn't work out nearly as well as I had once hoped. Despite those imperfect projects, all individually and collectively provided formative learning experiences that I now greatly value.

I continue to enjoy satisfaction in the quiet privacy of my own good company in pursuit of an elusive idea—contemplating new information and insights and writing, drawing, making or refining something in a tangible form that previously existed only in thought. It is also quite wonderful to engage together in joint projects with others who contribute special talents to accomplish common goals.

Viewed from this perspective, life by design affords a very personal, sometimes shared, multi-dimensional design challenge. And just as there are no universal solutions that apply to each of us, there are also no common one-size-fits-all ideals regarding what constitutes personal success.

I could just as appropriately refer to living as an "art" rather than apply the term "design." And although believing both to be equally important, there is an inherent sliding scale of personal art vs. societal design priorities I consciously apply to my various professional pursuits.

There are frequent circumstances when a project or task demands strict fealty to responsibilities dictated by concern for the safety, wellbeing and other needs of the client and broader public. My overarching priority as a designer in these instances is to fulfill and exceed externally-defined functional, practical, economic and yes, aesthetic prerequisites.

On other occasions I grant myself full liberty and license to pursue exploratory excursions into realms and endeavors which, although certainly no less challenging, are deliciously and unrepentantly self-indulgent. Although it's nice when others appreciate the results, I am accountable only to myself as the ultimate critic. It either accomplishes what I originally set out to do…or it falls short and I learn from the experience.

In all cases, as rich and satisfying as my life has been—still is—it is not my intent to present it as a model for anyone else. Each person must create their own joys of exploration, discovery, innovation and persistent survival that make mortal existence worthwhile and enriching.

Nor is it my intent to offer this publication as a retrospective catalog of my projects and accomplishments. At the time of this writing, I'm not quite done or dead yet.

None of this, of course, will discourage me in the least from telling you what I think is most important. This is an ordained right that comes with my long-time official academic appointment and office walls covered with various important-appearing professional certificates and honoraria. Besides, I didn't make you buy this self-indulgent book.

Hopefully, some important lessons that I have learned during the process of designing and experiencing my life will offer useful considerations as you continue to design and experience yours. I'll begin this narrative where I left off with some life lesson conclusions in *Thinking Whole*:

Lesson One is that if you can't sit still, then really get moving! You can't rely entirely upon the swift current of events to carry you safely through all of the obstacles. Instead, you sometimes have to paddle like hell to take command of your options so that they aren't left to chance or fall under the control of others.

People who constantly operate in a reactive mode are

typically overwhelmed, confused and ineffective. Even though you may not have all of the desired knowledge and confidence, it is important to set some form of purposeful, yet flexible plan into action. Put yourself in charge of your own choices. They are too vital to delegate.

Lesson Two is to make it a practice to look at the *BIG PICTURE* so that you don't get lost in the details. In your personal life, try to be clear about what it is that you care most about. In a new job or position assignment, work to understand the broad priorities and scope of the business or service environment of the organization, and the ways that your designated role fits into and contributes to them.

Constantly strive to visualize your participation within the context of larger purposes and processes which will be impacted by your performance. Focus on things that are most important and make certain that they are accomplished well and on time. Get and stay organized.

Lesson Three is to define your opportunities, responsibilities and prerogatives as ambitiously as possible. Push the boundaries of your personal expectations and job descriptions to the limit. Be more than anyone expects you to be.

Don't worry about what others might consider to be either beneath or above your appointed station. Find ways to help friends and associates look good and be an advocate for them. Think of any title or status that you have as an opportunity to take action rather than as a license to take credit. See yourself as a leader and act like one.

Lesson Four is to value your own resources and abilities. If you don't know how to solve a problem someone else's way, then devise your own approach. Don't allow yourself to be intimidated by things that you don't immediately comprehend merely based on a lack of background information and experience. Just absorb what you can and make it a point to later investigate points that you missed.

Learn to be patient with yourself. Don't be afraid to ask questions, but do this thoughtfully, especially when formal occasions warrant some discretion.

Heed Henry Ford's advice:

> *Whether you believe you can do a thing or not, you are right.*

Lesson Five is to learn from others, but to be yourself. Everyone has their own unique strengths and styles. When you attempt to emulate others or compete with them on their terms, you are likely to fail. Doing that often tends to cloud your awareness of your own unique qualities, impairs natural creativity and keeps you off balance. Besides, being who you are can be a lot of fun once you get the hang of it.

Author Background

LARRY BELL IS an endowed professor of space architecture at the University of Houston where he founded the Sasakawa International Center for Space Architecture (SICSA) and its space architecture and aerospace engineering/space architecture graduate programs. Prior to moving to Houston, Larry headed the industrial design graduate program at the University of Illinois. He was also a licensed practicing architect and senior associate with a leading regional firm.

Separate from his present academic career, Professor Bell has co-founded several private hi-tech companies, including one that grew through mergers and acquisitions to employ more than 8,000 full-time professionals, went public on the New York Stock Exchange, and was purchased by General Dynamics. He also served as lead planner for a national crime prevention through environmental design demonstration program sponsored by the U.S. Department of Justice in Washington, DC.

Larry is a former U.S. Air Force air traffic controller, a national award-winning patented inventor, and also a serious wood and stone sculptor who has exhibited his work in major one-person and group museum and gallery shows. In addition, he has published more than 1,000 opinion article columns on a wide variety of topics in Forbes, Newsmax and other national publications along with 11 books including one co-authored with

Apollo 11/Gemini 12 astronaut Buzz Aldrin.

Professor Bell's numerous professional honors include certificates of appreciation from NASA Headquarters, the Kyushu Sangyo University (Japan) *Space Pioneer Award,* and two of the highest awards granted by Russia's most prestigious aerospace society for his international space development contributions. His name was placed in large letters on the outside of the Russian Proton rocket that launched the first crew to the International Space Station.

Bell is an emeritus fellow with the prestigious invitation-only Explorer's Club, an associate fellow with the American Institute of Aeronautics and Astronautics (AIAA), and a member of the American Society of Civil Engineers (ASCE).

Larry Bell

Chapter 1: Personal Foundations And References

I RECENTLY HAPPENED upon an old lecture I presented more than a year ago which began with an anecdote which was clearly intended to warm up the audience. Although I don't remember if this transparent ploy worked, I'll begin this book by trying it out again on you.

As that story goes, an artist and an engineer went on a camping trip and pitched tent in the Texas Hill Country.

In the middle of the night, the engineer nudged his friend and asked, "Do you notice anything?" The artist answered, "That's incredible!—All those bright stars, each a sun probably much like our own, with planets. Maybe many of those planets have life, possibly even beings somewhat like us."

The engineer responded, "That's not what I mean. Did you notice that someone stole our tent?"

An intriguing question in your own response to this story is which of the two characters, if either, do you most identify with? Personally, I see myself very much reflected in both.

My previous book, *Thinking Whole*, envisions design as an activity that combines left brain analytical pragmatism and right brain creative vision. Here, design is a holistic endeavor that connects who we are, what we value, and what we have learned through our experiences. It requires that we draw upon all of our resources gained through personal and professional lessons to pursue proactive initiatives.

It is important to realize that personal limitations and professional boundaries that others encourage us to accept are not natural to our holistic potentials. Professions are typically established to protect "turf," not to empower comprehensive or visionary thinking. The true nature of design can be revealed by observing design in nature; natural design is not bounded by discipline-based categories we naively adopt.

As with all our life experiences, lessons taken from natural systems and processes are transferable provided that we seek ways to apply them.

Design fundamentally involves deliberate choices. These decisions include the personal goals we select and needs other than only our own we recognize and respond to. Other choices involve possibilities we have the vision and courage to pursue, and the full range of personal resources we are prepared to develop and utilize.

My design (and art) interests are broad and diverse, and I wouldn't want it any other way. Some activities are directed to very personal self-defined goals and objectives. Others respond to externally-dictated requirements that are often driven by very specific, difficult and complex conditions. This second category demands careful consideration of many issues, including costs and means of production, market demands and user needs, human safety and performance concerns and product or system implementation and operational factors.

When we are designing something, it is extremely important to recognize where exactly our accountabilities lie: whether within ourselves as with fine arts; or to others and the environments that our works will impact.

Consider who is bearing the costs and risks, whether the project is something really worthwhile, and what planning and capabilities must be brought to an endeavor to produce a successful outcome for everyone.

I clearly view entrepreneurship as a design activity that can create high-value capabilities and products. There can be no doubt that it embodies a need for vision, innovation and market savviness that characterizes much of what designers do.

Through design we recognize goals that reflect what we and others truly care about; we formulate plans with reasoned purposes and approaches; we conceive and evaluate options which respond to opportunities and needs; we visualize and describe ideas and solutions that can be acted upon; and we undertake projects that transform possibilities into realities.

Design is not only what we do as an activity or service, or what we create as products of our efforts. Design is also a pursuit that can lead us to new adventures that bring added meaning, satisfaction and purpose to our existence. In this sense, design is an intellectual and spiritual endeavor that is most fully practiced as an integral and fundamental aspect of life.

While some of the art and design activities I will discuss have involved only me, many others have been undertaken by teams of people from organizations I have created, co-founded and/or headed over the long span of a half-century. These programs and projects address many topics, ranging from sculpture on the most personally-accountable end of the spectrum, to transportation design and architecture for space and other extreme environments on the externally-driven end of the scale.

Some of these endeavors were undertaken by entrepreneurial commercial entities that I created with partners. The largest of these, an aerospace company, ultimately grew through mergers and acquisitions to employ more than 8,000 professionals. A couple of others had between 30-35 people.

I do not feel qualified or motivated to offer design theories or mantras that can be generalized as universal truths. Instead, I will share some examples of design experiences and products that reflect what design has meant in the context of my life. I hope that they will have some relevance to yours as well.

Reflections on My Life

My life, by design, has exposed me to countless adventures of body, mind and friendship, and continues to do so today. It has led me from the small Midwestern town of my childhood to exciting places and experiences throughout the world. And it has introduced me to wonderful people, ideas and possibilities what have enriched my world.

Most of these developments and encounters were never planned, or at least the outcomes seldom were. Instead, they usually resulted as consequences of pursuing personal interests...interests that were unbounded by professional categories or expectations espoused by others. My life fits me, just as it should. After all, I designed it.

We don't design our lives all at once like a product. Life isn't a product at all...but rather, life is the source of our creations. Life is something we design moment by moment, interest by interest, decision by decision and resulting opportunity by opportunity.

I have learned that sometimes those opportunities speak in quiet voices, including our own, that we have the wisdom to heed. Sometimes they arrive in the form of problems and challenges we recognize and are willing to accept. And sometimes they grow out of seemingly unfortunate conditions that force us to seek and pursue alternate pathways. Design characterizes the way we understand and respond to these opportunities, and the process keeps us moving forward.

Each of us has values, expectations and perspectives that are influenced by our cultural and family backgrounds. These formative experiences and our adaptive responses to them offer foundations and references that should not be ignored.

What we are now has a genesis that began when we first entered this world, and perhaps even before. Proactive design commenced when we learned we could influence the movements of our limbs and lungs, along with reactive attentions of those around us.

It has become repeatedly apparent that many of my present interests and pursuits can be traced back to much earlier exposures and experiences, although not always in a direct or continuously linked chain of occurrences. Maybe you will discover similar patterns in your life.

My journey of life by design began in small town settings. Although I now reside in the city of Houston, my small-town beginnings continue to be an enduring part of who I am. It was in those settings that I first discovered joys and challenges of friendship, flirted with romance, acquired formative lessons and values, was first exposed to responsibility, and enjoyed fun and adventures which were not all known to my parents.

Childhood through high school years in that small-town Midwestern environment were wonderfully wholesome and enjoyable, with a bit of adventure thrown in to boot. As teenagers, and entirely unbeknownst to our parents, a pal and I built a canoe out of wood and canvas, then set off on the Wisconsin River for its connection with the Mississippi about 90 miles away. Upon arrival, we left the slightly leaky craft at Prairie du Chien, Wisconsin, and hitchhiked to New Orleans.

Our parents gratefully welcomed two very sunburn-blistered, mosquito-bitten and hungry kids home after we had panhandled enough money to purchase bus tickets back. Our escapade even made us heroes among classmates on the front page of the local newspaper.

My parents established and operated an airport in Baraboo, Wisconsin where I spent early years around private airplanes which included experimental craft constructed by owners from plans and parts kits. That background has likely influenced my subsequent interests in aerospace design.

Dad, a flight instructor and charter pilot, gave me flying lessons. With his hands and mine both on the dual controls, our mental and muscular commands seemed to merge in a way that made it impossible for me to be certain which of us was really responsible for causing the aircraft to bank and turn or climb and descend.

One day after we had landed, he unexpectedly suggested that I take my first solo flight. Before I fully realized what had happened, I found myself looking down upon the miniaturized landscape below and suddenly became conscious that I was more alone than I had ever been in my life. It was immediately clear that I was the one doing all of the flying, and that I would have to be the one to do all of the landing as well.

At that point I experienced very mixed emotions. It was exhilarating to enjoy new sensations of freedom and independence as the *140 Cessna* responded to my wishes on that beautiful summer afternoon. Yet, I also felt considerable apprehension that maybe I wasn't quite ready for the challenge of returning to the security that I had taken for granted only a few minutes earlier. There was no longer anyone else who could help if I needed it, and no option except to succeed.

The landing wasn't as smooth and easy as I had wished. I circled the pattern and lined up for my final approach at the proper distance and altitude according to plan, but my airspeed was excessive. As a result, I hit the surface pretty hard and fast, bounced high, and could see that I was rapidly running out of runway.

Finding myself airborne again following my brief contact with terra firma, I advanced the throttle and pulled back on the controls to go around for another try. In my determination to do this I reacted too aggressively, nearly causing the plane to stall out and execute an unscheduled encounter with the tree line ahead.

That wonderfully resolute craft was not about to surrender its own future to my incompetence and pulled us both back towards the clouds. Grateful for another chance to get it right, I circled the field again, landed without incident, and apologized to my tolerant host.

After the Cessna and I both arrived back, each in one piece, I asked dad if he thought I should go up for another spin. He smiled somewhat painfully and suggested that maybe I had experienced enough excitement for that day. I am sure that he was actually speaking for himself as well.

Of special personal importance, the private airport that my parents built and operated there still exists, now as an active municipal facility. A circular paved and landscaped memorial site that I created for them originally contained two sentinel trees to symbolize my parents still together, standing side-by-side. Though those trees are now casualties of harsh winters, the earth they stood on berms up to the top of a curved retaining wall which forms a bench overlooking a runway; It is a place where people can rest and watch airplanes. Mom's and dad's ashes are buried there, and I feel comforted and close to them when I visit that place.

I now must come to terms with the area I remember having changed in many ways that are difficult to adjust to. Just up the highway from the airport where I spent many youthful years, a large indigenous Indian-owned and operated bingo gambling casino has been built on formerly rural land. The place is packed with people throughout the day and night and has become a major employment center where small town residents work among bow-tied blackjack dealers and net-stockinged waitresses.

The near-by state park where we roamed freely and partied boisterously at all hours is now crowded with tourists. The private cottages it once had are gone, and it is now publicly controlled with tightly regimented rules, including closing hours. I doubt that many local residents go there anymore.

Fast food franchises and big commercial outlets have

drained business away from the friendly mom-and-pop restaurants and stores that I remember, and most of them have disappeared. So much for nostalgia!

Maybe it's a blessing that we really can't go back to recapture our pasts. After all, what's so bad about the present?

As wonderful as those memories are, they don't offer the vitality and excitement of anticipation that we experience in our current lives. The past is like a marvelous book we have already read. We have enjoyed it and learned from it. And perhaps some new discoveries are revealed when we read it again.

One of those lessons is probably that we can't ever go back. Another is that *home* is really where we live among people who still need us and events that we are still part of.

Adapting to Time Changes

Those were very different times than today. For example, I didn't grow up with any notion that guns are bad, now a prevalent view evidenced by more recent generations. This view is held most particularly among individuals who live in large metropolitan centers, never served in the military, and frankly, have little interest in or knowledge of American history.

I grew up with guns, and although not a hunter, still actively enjoy shooting them. Doing so requires me to concentrate full attention upon a target goal, testing my ability to respect the remarkable precision of a carefully tuned completion weapon.

My first gun was a Daisy Red Ryder BB gun costing $4.95—likely comparable to more than ten times that amount in today's money. I worked to earn that hefty sum mowing neighbors' lawns in spring and summer, raking leaves in the fall, and shoveling sidewalks in winter. Within a year I had managed to save about half of that amount, and expected that I would have enough to complete the deal in another.

A miracle occurred in the form of a grandpa who came to visit. I couldn't figure out why, but he suggested that we walk to town together.

Grandpa stopped right in front of the very same hardware store where that gorgeous BB gun—the one with the real wood stock—was displayed in the window. When he commented about how nice it was, I sadly agreed, but told him that it would be quite a while before I could afford it.

Nevertheless, grandpa suggested that we go in and take a closer look anyway. When we left, if you can possibly believe it, I owned that wonderful boyhood treasure long before I ever expected. I was a very lucky and happy kid for sure.

As children, my pals and I spent many afternoons watching Western gun toting movie theater matinee heroes win over the bank robbers and cattle thieves. We could always tell who the good guys were. They were the ones with the clean white hats, biggest horses, and who invariably won the hearts of the best-looking gals.

It possibly helped in the female adoration that our aspirational role models never seemed to sweat much during hot desert bandit chases, and that some of them, Roy Rogers and Gene Autry for example, even sang to them pretty well.

Overall, these characters were straight shooters in more ways than one. They didn't talk dirty, smoke dope, take off even so much as their boots, or do much of anything else that attracts audiences today. Yeah, they shot or beat up lots of those bad guys, including native Indians who were all-to-often savagely caricatured. And sure, the scripts were simplistic with predictable plots and beautiful sunset endings.

Since then we have witnessed the popularization of anti-heroes who help us make peace or war with our subsequent loss of innocence. Lead figures can be alcoholics, drug users and criminal types that evoke empathy because they are presented to have experienced hardships and express sensitive, humorous or other redeeming human qualities.

We watch them as they return bitter and confused from senseless wars; as they make serious mistakes and hurt themselves and others; as they engage in sexually explicit, yet meaningless relationships; as they encounter and react to bigotry; and as they voluntarily—or not—live or work under depressing and hopeless conditions.

Glamour is often associated with power and wealth achieved at the expense of ethical principles. Killing and suffering are made exciting, yet mind-numbing, through the use of special effects, slow motion and close-ups.

Blessedly, there are also many kind, positive and beautiful movies that capture better aspects of our nature. Less fortunately, we may often have to go to specialty theaters featuring foreign films with subtitles to see them.

Maybe it is sometimes good for us to see movies that depict the seamy side of life and highlight human frailties, if only to illustrate how vulnerable we can become if we're not careful. Perhaps they also remind us that life can still be worthwhile even when conditions that surround us are far from perfect. And possibly, it is constructive to put some of those old romantic fantasies behind us and boldly look reality in the eye with appreciation for opportunities bestowed upon us.

But isn't it fun to occasionally return to those simpler and more idealistic attitudes and experiences of the past? To watch actors and actresses we have admired, now long gone, become young, hopeful and determined again? Isn't that okay once in a while?

As for today, after I finish writing this pathetically nostalgic reflection I just might see if there's a re-run of one of the old wild-West movies on TV. Although it's not so easy to find anything with Lash La Rue, Hopalong Cassidy, or even Randolph Scott or Gary Cooper anymore, maybe there's something featuring John Wayne.

My wife Nancy will probably give me hard time for watching anything so mindless and trite. But then, how can she possibly understand? She's originally from Chicago.

Later in life I have become somewhat of a history buff on the American Civil War, a brutally tragic time of human sacrifice and valor. Letters from leaders and foot soldiers to loved ones reveal understated, stoic, poignant, thoughtful, sometimes humorous marvels of human courage and character under horrific hardships and unimaginable suffering. The intelligence and self-educated literacy of the authors offer remarkable and ageless lessons.

Introductions to Responsibility

Back in the mid-1950s I moved to Milwaukee the day after high school to work for the Inland Steel Corporation as a floor panel fabrication designer for commercial buildings. Upon recommendations from my teachers, I was one among six the company had pre-selected from state schools for this opportunity as an experiment in hiring and training non-college engineering degree-holding candidates.

It soon occurred to me that although the new position was quite an honor and responsibility for such a young guy, it also portended a dead end, albeit in a corner office with a better view. Not willing to imagine that place as a life destination, I decided to escape before career inertia inflicted irreversible mind-control.

But there was a big problem.

While still in high school, I had joined the U.S. Navy Reserves, and had even gone through boot camp training at the Great Lakes Naval Training Center. Under the enlistment terms I was obligated to enter active duty within a six-month period which was scheduled to lapse soon after I had taken the position with Inland Steel. The influential company negotiated with the Navy Department that my active service obligation be delayed for two years because they asserted that my work for them was related to national defense.

The deal also provided that I would then serve the government as a naval designer for two years and serve out two more working for Inland after that. In other words, all options for an elective career path were rapidly closing. There was only one exit door—and I took it.

I was allowed to transfer from reserve status of one military branch to active duty in another. With that option available, I enlisted in the U.S. Air Force. In retrospect, that highly instinctive, seemingly impulsive decision is one I have never regretted.

Top-tier overall scores on Air Force post-enlistment aptitude tests catapulted me, then a fresh-out-of-high school 18-year-old, into challenging new experiences of awesome responsibility as an air traffic controller—more specifically as a ground control approach (GCA) operator. Upon completing basic training at Lackland Air Force Base, in San Antonio, Texas, I was transferred to Keesler Air Force Base, Biloxi, Mississippi for six months of initial GCA training.

GCA, which entailed "talking in" aircraft using precision radar under low visibility conditions, was somewhat like flying multiple aircraft at once, whereby all relied upon your guidance as they flew around the "airport pattern" in preparation for landing. Then, on the "final approach," you constantly communicated precise glide path, runway alignment and other instructions to each individual piloted craft until their wheels touched down on the runway.

Those pilots depended on us to respond quickly and appropriately to all sorts of routine and contingency circumstances.

Our confident voices and error-intolerant instructions were products of expansive training and testing. We learned comprehensive air traffic control and airfield rules, technical details about the radar systems, how to regularly check and recalibrate scope accuracy adjustments in the dark using small screwdrivers and how to continuously servo radar antennas towards incoming aircraft with foot pedals while simultaneously manipulating display and communication controls with our hands while all the time issuing pilot guidance.

We practiced memorizing and keeping track of multiple individual aircraft call sign identities, each correlated with types, altitudes, approximate airspeeds and headings, which appeared only as undifferentiated moving "blips" on our scopes. This training involved endless trial runs vectoring diverse mixes of live aircraft, including fast fighter jets, lumbering cargo carriers and prankster-prone helicopters.

Our instructors and co-conspiratorial pilots concocted a fiendish variety of faux emergency situations to challenge our problem-solving skills and composure under pressure. The final test was a doozy: a Master Sergeant yelling in my ear that I was doing everything wrong as I talked in a jet pilot claiming to have a landing emergency with a lost engine and erratic control under strong fluctuating crosswinds.

Deciding that I couldn't allow the loud harangue behind me to distract full forward concentration on the challenge at hand, I tuned it out in my mind. As I later learned, that was exactly what I was supposed to do. It was considered to be the most crucial aspect of my ultimately successful assessment.

The ensuing years as an air traffic controller were among the most satisfying, if also most demanding, professional responsibilities I have ever known. It was enormously gratifying to open the radar unit door, see the foggy ghost of a large aircraft on the nearby runway, and hear the grateful voice of its pilot on your headset say, "Thank you GCA...great job!"

My tour of military duty tour concluded with a fourth and final year-long posting at Sondrestromfjord, Greenland, a small isolated base with a single one-way in and out landing-takeoff strip closely flanked by two tall mountains and terminated by another. Inclement weather was frequent, and the mountains blocked radio and radar contact with aircraft that strayed off course.

Yes, I acknowledge that I'm waxing nostalgic about those often fabled "good old days" before superior automated computer-based systems replaced antiquated technologies of my "back-in-the-day" relic generation. Yet I can't help but doubt that these marvelous new contrivances have made the urgency of keen human judgment obsolete as well.

I continue to believe that vetting of the very finest candidates still really matters. Among these, my former Chanute Air Force Base air traffic control team mate, life role model, professional mentor and close friend Staff Sgt. Curd, a black man from Alabama, was a personal all-time favorite.

Charlie, if you're still out there somewhere, thank you for your personal GCA guidance…great job!

Warm Memories from Greenland

Sondrestromfjord (or as we called it, *"Sondey"*, or *"The Rock"*) was primarily an emergency refueling base for a much larger and more active Thule Air Force Base which served as a radar surveillance site during the Cold War with the Soviet Union. Located near the edge of the ice cap about 90 miles inland from the Davis Straits, there was no soil on the rock surface to support trees or other vegetation.

Temperatures during the long Arctic night sometimes dropped to -60 degrees Fahrenheit, and frequent high wind conditions made the effects of the cold even more penetrating…we were virtual shut-ins. This circumstance was particularly depressing during our several-day-long work shifts at the radar site location which was remote from the main base of about 100 airmen.

There, at the radar site, our two-man teams were confined in a tiny one-room standby shack while waiting to assist inbound aircraft. Our only other company was an occasional Arctic fox that observed us in the dark as we waded through drifted snow to and from the radar unit—or, since we had no plumbing, as we went out to answer nature's call. The latter mission had to be performed quickly in peril of frostbite injuries which would cause sitting to be painful, or potentially impact any future family plans.

Lacking normal outlets for recreation, we spent much of our excess free time engaged in ping pong, card games, recorded music and some Japanese *Godzilla* movies that had been brought in by someone with either bad taste or good humor. Mail was infrequent, about every month or so, and I regret to say that morale was generally very low.

I have since come to know Arctic and Antarctic explorers and scientists who have happily endured incomparably worse conditions, but the difference was that they did so to achieve important personal goals. Most of us felt that we were wasting a year of our lives cut off from everything we really cared about for no reasons that we could understand. Some fellows psychologically dropped off the deep end of depression, and to a man, we all had extremely negative opinions of that place.

Then, one day, I hit upon the idea of getting into the portrait painting business. Since we didn't have commercial services there to rely upon, just about everyone took up an avocation to help earn a few extra dollars. (My first experiment as a barber was a failure that left my victim nearly bald.)

Using pigments left over from scavenged paint-by-number kits and scraps of tarpaulin stretched over plywood for improvised canvasses, I began to produce paintings of friends' loved ones from photographs that they supplied. Before long, I found myself sitting outside for hours on end, fingers numb with cold, trying to capture the magnificence of those mountains that I had previously cursed.

That landscape, which had seemed so barren and brutal,

began to reveal itself in a whole new way. Its scale and rugged beauty demanded my respect, and that recognition illuminated something inside me.

I realized that there was a very good reason to be there after all. What better place to discover the excitement of my own fragile existence?

I learned important lessons during that unplanned year in Greenland. One is that changes that are forced upon us can provide unique opportunities to discover unexpected aspects of who we are. Another is that the wonderment we often fail to see around us may lie hidden behind mental blocks of ice and mountains of stone.

Lessons from Dumb Mistakes

Truth be known, I had definitely not volunteered to spend the final twelve months of my four-year U.S. Air Force tour of duty in Greenland. As someone who values mild climates and warm female companionship, it was one of the last places I would have chosen. Unfortunately, neither the U.S. Department of Defense nor Father Fate had solicited my preferences in this matter. No, a colonel whose authority I made an enormous error in challenging made that decision entirely without any consultation on my part.

My dumb judgment misadventure began when our squadron headquarters was being relocated to another on-base facility. Everyone pitched in to help the move, and I voluntarily contributed by painting various signs for the new building and parking lot.

The graphic quality of my work drew admiration, and it wasn't long before other squadron units asked me to paint signs for them. I obliged good heartedly—at least for a while, until I began to tire of this charity effort.

One morning, I was awakened by a telephone call from a Colonel who, in a most pleasant manner, asked if I would be willing to paint a large sign to be placed in front of our regional headquarters, which was also located on the base. I politely declined, reminding him that as an air traffic controller, sign painting wasn't part of my job description. Doing so would also exceed the number of hours I was allowed to work in my safety-critical field according to regulations.

The biggest part of that mistake was informing him that while he could order me to do so, it would compel me to file a formal protest.

A short time later, I received transfer orders which reassigned me to the Sondrestromfjord Air Force Base in Greenland. As it turned out, that colonel was in charge of staffing overseas bases.

This experience turned out to be a teachable moment that has lasted a lifetime. Namely, it is that when someone in a superior position asks you to do something, it's a good idea to check their area of authority before saying no.

Another formative lesson was to never again allow myself to be in a position where anyone would be able to send me involuntarily to Greenland—or to anywhere else. Concluding that such independence would require having money, I enrolled in the economics curriculum at the University of Wisconsin—Milwaukee to learn how to earn it.

Father, Lloyd

Mother, Evelyn

1st home near dad's gas station

Me (L) with cousins Judy & David

Me (L) with David & sister Lorna

My early out-of-box experience

I wanted to be a cowboy

My first "chick magnet"

Happy day with BB gun and friend

Solo day at dad's airport

Bell airport in Baraboo, Wisconsin

US Air Force days: Air traffic control with Sgt. Charlie Curd

GCA radar unit at Chanute Air Force Base, Rantoul, Illinois

Sondrestromfjord, Greenland: "The Rock"

Larry Bell

Chapter 2: Going Academic

UWM WAS A very small campus at that time, with little social life to distract me from my mercenary goal. Having already spent four years in the military I felt that there was little productive lifetime left to waste, plus my existing money supply tank was running on empty. All proceeds from selling my Greenland paintings went to covering a single semester of tuition.

I addressed the money problem by posting notices in the Milwaukee Journal want ads advertising architectural drafting services. A homebuilder responded, and I began designing for his residential clients. In addition, I get free rent along with some meals in exchange for cleaning up a restaurant located directly adjacent to the UWM campus which was owned by my very nice landlady.

As for learning how to make lots of money, that wasn't nearly as clear and immediate as I had hoped. The emphases upon macro and micro economics, bull and bear markets, and public policy administration weren't going to put money in my empty pockets any time soon.

Worse, I discovered that I didn't even care about that purpose very much. It was more important to do something that was creative and useful...like, for example, the home design I was already doing.

So a year later I transferred to the five-year-long College of Architecture program at the University of Illinois. Together with Sally, a lovely girl I met on my first day UWM English and married that following summer, we left for Champaign-Urbana.

That move commenced an expansive series of episodes and byproducts in my life by design. And yes, some finally even made quite a lot of money after all.

A Fountainhead of Ideas and Ideals

Those undergraduate architecture years were financially challenging and meager times. Sally, who had dropped out of school to join me, got a job as a service representative, and I scored an opportunity to work a heavy part-time schedule with the University of Illinois Architects' office. I supplemented that income selling blood at two different hospitals to get around restrictions limiting the practice to once per month. One hospital paid $25 per pint; the other $15.

Our credit was good enough to enable time payments on a used 8 ft. by 40 ft. house trailer that we lived in over the next several years. Many of the other residents in the park were married U of I graduate students about the same age I was—couples living on frugal budgets as we were. None that I can remember felt "poor," because these conditions were elective and temporary as they pursued dream careers.

Our neighbor community represented a diverse and interesting mix of professional interests. My good friend Merrill Garret went on to continue his studies of dolphin communication as a professor at MIT.

My commitment to the five-year architecture program was a very serious matter. Having enrolled later than my fellow students I felt personal pressure to recover lost time. Adding to this, Sally and I were both heavily invested in my academic performance under severe study time constraints imposed by outside work necessities.

The College of Architecture admission officer had discouraged me from even trying to enroll in the demanding, time-intensive curriculum while also working part-time. He predicted that my chances of success were quite low...later congratulating me for not taking his advice. That occurred after I passed the oral exam he presided over as part of the State of Illinois architecture licensing accreditation and National Council of Architectural Registration Board (NCARB) requirements following my fulfillment of college graduation and internship prerequisites.

That original decision to enroll in the U of I architecture program opened my awareness to a brave new world of ideas and ideals that had previously been unimaginable. It was a domain inhabited by towering giants of flesh and granite, a vast landscape of visions and deep oceans of unrealized possibilities.

As now a relatively young architecture student from a small Midwestern town, I immediately became awed to realize that I had stumbled into an exciting new world dominated by giant spirits—deities whose eternal presence scorned any compromise of authentic virtues.

This was, above all else, the era of Ayn Rand's *The Fountainhead* and Howard Roark, implicitly representing a superhero version of architect Frank Lloyd Wright battling valiantly against forces of mediocrity. As Lebbeus Woods, my classmate and closest friend at the time observed many years later, the story "has had an immense impact on the public perception of architects and architecture, and also on architects themselves, for better or worse." [1]

There can be no doubt that Ayn Rand's epic story continued to influence perceptions of architects and the

---

[1] *The Fountainhead: Everything that's Wrong with Architecture*, Lance Hosey, Architecture Daily, November 14, 2013.

profession long after it was first published as a book in 1943 followed by a movie version starring actor Gary Cooper a few years later. I would only modify Leb's observation to describe that impact as being for better *and* for worse.

On the "for better" side, Rand's portrayal of ideal human virtues inspires high moral values and professional purposes. In a 1932 letter to actor Colin Clive, she wrote:

> *There is nothing to approach the sanctity of the highest type of man possible and there is nothing that gives me the reverent feeling, the feeling when one's spirit wants to kneel, bareheaded. Do not call it hero-worship, because it is more than that. It is a kind of strange and improbable white heat where admiration becomes religion, and religion becomes philosophy, and philosophy—the whole of one's life.*[2]

Ayn Rand thought she had found the essence of the ideal man in Wright. In a letter to him seeking an interview, she wrote:

> *[The story of human integrity] is what you have lived. And to my knowledge, you are the only one among the men of this century who has lived it. I am writing about a thing impossible these days. You are the only man in whom it is possible and real. It is not anything definite or tangible that I want from an interview with you. It is only the inspiration of seeing before me a living*

---

[2] *Letters of Ayn Rand*, ed. Michael S. Berliner, New York: Dutton, 1995, p. 16.

*miracle—because the man I am writing about
is a miracle I want to make alive.*[3]

Rand had conceived Roark to champion free-thinking self-reliant virtues which were antithetical to the oppressive Soviet communist society her family fled in 1926. She imbued his character with qualities she most admired: to think independently in forming values and judgments, and to live with complete integrity—to be true to self.

Being true to self, in turn, requires that one must have a self—a self that can recognize its individual responsibility and power to make choices that will control the outcomes of its life. Integrity, therefore, embodies a commitment to action, the "practice what you preach" virtue.

Rand rejected the notion that to succeed, one must "play the game," or conform to practices of one's company or profession if one finds them unethical. She exhorted that successful living forbids a person from betraying their mind, whereby moral and intellectual virtues are requirements for practical success, not hindrances.

Roark, the architect, epitomized Rand's ideal role model as a fiercely independent creative force who believed completely in the merit of his revolutionary designs. At one stage, despite being financially destitute, he gives up a lucrative, publicity-generating commission in order to stand by the integrity of his design. He takes a laborer's job in a granite quarry rather than compromise a detail in one of his buildings.

Roark's idealism—which some might view as an epitome of egotistical hubris—goes so far that he dynamites his own perfectly serviceable social housing project because he doesn't like the aesthetic. He simply explains: "I destroyed it because I

---

[3] *Letters of Ayn Rand*, ed. Michael S. Berliner, New York: Dutton, 1995, p. 109.

did not choose to let it exist."

I can't help but be reminded here that Frank Gehry, one of the most famous architects of our time, has expressed similar self-ordained Roarkian authority. He said:

> *To deny the validity of self-expression is akin to not believing in democracy—it's a basic value—if you believe in democracy then you must allow for personal expression.*

Yet as Paul Davies observes in an *Architectural Review* article,

> *...democracy is the will of the majority, not the individual, and Ayn Rand hated democracy because she felt that it crushes personal freedom. When the lines between individualism and democracy blur, it's safe to say that Rand's ghost still haunts us.*[4]

As with Gehry's designs, many contractors and fabricators would have nothing to do with Wright's projects due to their construction difficulties. Wright constantly experimented with new materials, techniques and details which added complexity and expense. In addition, they often didn't weather and age well.

Having said this, I believe that Rand's philosophical legacy projected through Howard Roark and channeled through Frank Lloyd Wright's aesthetic sensibility warrant enduring respect.

In Wright's autobiography, he writes:

[4] The Architectural Review, *Howard Roark (1928-1943)*, December 5, 2013, Paul Davies

I knew well by now that no house should ever be on a hill or anything. It should be of the hill, belonging to it, so that hill and house can live together each happier for the other…The lines of the hills were the lines of the roofs. The slopes of the hills their slopes, the plastered surfaces of the light wood-walls, set back into shade beneath broad eaves, were like the flat stretches of sand in the river below and the same color, for that is where the material that covered them came from.[5]

Rand's description of Roark's aesthetic leaves no ambiguity regarding Wright as its source of inspiration:

It was as if the buildings sprung from the earth and from some living force, complete… unalterably right…Not a line seemed superfluous, not a needed plane was missing…He had designed [them] as an exercise he had given himself, apart from his schoolwork; he did that often when he found some particular site and stopped before it to think of what building it should bear.[6]

As a parallel reflection, I hold similar admiration for works of Victor Lundy, another architect and former colleague of mine at the University of Houston whose works evoke that same essential natural quality—a sense of perfectly belonging in their environment—a timeless feeling of inevitability that it

---

[5] *An Autobiography*, Frank Lloyd Wright, Vol. 2, Collected Writings, New York: Rizzoli, 1992, pp. 224-27.

[6] *The Fountainhead*, Ayn Rand, New York: Scribner, 1986. P. 7.

was always meant to be there, and always should be. I have referenced Victor's projects in my previous book, *Thinking Whole*, as testaments to our marvelous human powers to transform visions into realities.

In addition to Wright, Rand's *Fountainhead* also portrays a thinly disguised version of his former "Lieber Meister" employer Louis Sullivan who appears in the story as Roark's mentor named Cameron. Just as Sullivan introduced the term "skyscraper" into the urban lexicon, Rand's fictional Cameron character realizes that a tall building should look tall—existing as a single emphatically vertical entity rather than as a stack of separate masonry structures.

Rand wrote:

> *There it was, in delicately penciled elevation. I stared at it and sensed what had happened. It was the Wainright Building—and there was the very first human expression of a tall steel office building as architecture. It was tall and consistently so—a unit, where all before had been one cornice building on top of another cornice building.*

And as Wright wrote:

> *Until Louis Sullivan showed the way tall buildings never had unity. They were built up in layers. They were all fighting tallness instead of accepting it. What unity those false masses that pile up toward the New York and Chicago sky have now is due to the master mind that first perceived the tall building as a*

harmonious unit.[7]

Wright was but one of many creative and philosophical giants we accorded great homage to. Other reverently honored, inspiring and critically judgmental spirits that inhabited our undergraduate design studios and hovered over our drawing boards included Ludwig Mies van der Rohe, Le Corbusier and Walter Gropius, who founded the Bauhaus. Multi-disciplined faculty of that world-renowned academy also included: architect Otto Bartning; artists Paul Klee, Johannes Itten and Wassily Kandinsky; painter-photographers Herbert Bayer and Lazlo Maholy-Nagy; and graphic designer Joseph Albers.

There were other iconic architects who particularly impressed me.

Louis Kahn's work expressed a monumental and monolithic style which combined modernism with the weight and dignity of ancient structures. His love affair with materials was reportedly communicated to his students at the University of Pennsylvania with the following dialog. He told them that if they were stuck for inspiration they should ask their materials for advice:

> *You say to a brick: 'what do you want, brick?'*
> *And the brick says to you: 'I like an arch.'*
> *And you say to the brick: 'Look, I want one*
> *too, but arches are expensive and I can use a*
> *concrete lintel.' And then you say: 'What do*
> *you think of that, brick?' Brick says: 'I like an*
> *arch.'*

Bucky Fuller's visionary inventiveness and pursuit of utmost

---

[7] *The H.C. Price Tower*, Frank Lloyd Wright, Architectural Record 119, Feb. 1956, pp. 153-160.

simplicity and efficiency of structures appealed to my desire for rationality, while at the same time, his endlessly long yet wonderful lectures I attended seemed to challenge any comprehension at all. That logical yet also mystical world he represented remains forever embedded in my psyche.

Felix Candella's elegant thin shell reinforced concrete structures seemed to disprove the law of gravity as only a myth. His mastery of the material combined a keen artistic visual intelligence and technical virtuoso understanding of engineering principles.

Candella sought to solve problems through the simplest, most direct means possible. His shell designs relied upon basic geometric properties of natural shapes, including domes and hyperbolic paraboloids, to eliminate efficient forms which avoided extraneous tensile forces—rather than through complex mathematical analyses.

I later had the great pleasure to spend a couple of days with Felix in Mexico to visit some of his projects. It is a treasured memory.

My early exposure to the teachings and examples of architects, engineers, artists, and philosophers revealed two seemingly parallel and independent, yet inevitably connected and interdependent worlds of discovery. One was exemplified in the domain of Bauhaus thinkers and makers who tended to most intently pursue spiritual and aesthetic visions. Other psyches dwelled more actively in regions governed by physical laws and social priorities.

I envisioned and continue to embrace architecture as the ultimate Gestalt—a tangible embodiment of ways humans and all other organisms shape and adapt to natural and social environments—a field of exploration and endeavor which encompasses a cognitive sum total far greater than the endless collection of individual issues and elements that comprise its parts—in short, unified and holistic thinking and action.

I don't recall ever imagining myself in the personage of a

Howard Roark figure standing defiantly in an eternal battle against compromise atop a Wainright building tower—or any tower, ivory or not. Nevertheless, I have ultimately discovered that there is much value to be found in Ayn Rand's emphasis upon self-reliant determinism.

Early failures to score high outside approval from professors on assigned design challenges was attributable to that precise reason—I was seeking it from the outside, rather than based upon a quest for authentic personal exploration and discovery.

In retrospect, that tendency was understandable. Attainment of top grades was a responsibility I owed to Sally as well as to myself to advance progress towards our shared future. Being risk-aversive, I sought to observe which design approaches and solutions earned winning top grades for other students—how I might find a formula of some kind in their successes that I might imitate.

Due to heavy outside work demands, I also had less time than did my academic peers to undertake those projects. This circumstance increased pressure through discipline and fatigue under famously tight assignment schedules in a round-the-clock architecture studio culture.

I never again received a mediocre grade after two important realizations finally sunk in which I have consistently applied ever since:

- Lessons one is to think hard and fast early in a project to avoid dead end ideas and wasted efforts.
- Lesson two is to always exercise authenticity. Don't merely try to second-guess what is wanted or emulate what others are doing in order to win approval.

My Temporarily-Delayed Genius Status

My personal Fountainhead-inspired "true-to-self" virtue was soon tested in for-real architectural practice. The firm, Richardson, Severns, Scheeler & Associates (RSSA), was headed by the late Ambrose Richardson, former lead designer for Skidmore, Owings and Merrill (SOM), one of the largest architectural practices in the nation.

"Am," as we all affectionately knew him, had established RSSA simultaneously upon his appointment as head of the University of Illinois' graduate architecture program. My initial fresh-out-of-college RSSA assignment there was to act as his de facto understudy and as a project design manager for a new $20 million dormitory complex at Eastern Illinois University. My assignment to the project—a very prestigious and large-budget project at that time—evidenced his confidence in me.

The practice at RSSA was to present three alternative design proposals for client discussion and ultimate selection. Although the various concepts might be either be prepared by the same or different architects, there was always one that represented the respective firm partner's top preference.

It didn't take me very long to observe an unspoken pattern in these presentations wherein the boss's choice always appeared as the third option in order to highlight features which made it superior to the other two. In the case of the university dormitory project, I developed one scheme entirely on my own, while Am and others developed the other two.

Feigning ignorance of the conventional set-up, I did what any self-determinate Ayn Rand acolyte would do—namely to arrange the following day's client presentation to make my scheme last. Whether or not that ploy influenced the decision I can't be certain, and I would like to think on the *not* side. In either case, the university officials unanimously chose mine.

And who could truly blame them? The dramatic twin 18-story triangular structures I proposed could hardly be compared with the rather bland low-rise complex which my reordering of the presentation options preempted.

I was soon to learn a painful lesson about inherent penalties of audacity. As I continued to work alone to further develop the selected scheme, office tensions ensued as a group of other architects continued to work on the internally preferred plan, which was to be presented again in hopes of client reconsideration.

During preparations for a fateful final concept selection presentation, one of the partners angrily chastened me quite loudly so that all could hear: "Larry, don't try to be a genius!"

The most humbling part of this episode occurred the following day when the client's asked the deciding question. Again reiterating their preference for my scheme, when they asked if it could be built within the available budget the reply was "no." Having run the comparative cost numbers, I silently disagreed.

Witnessing that second-rate concept win out for that realization constituted a very formative lesson in the politics of corporate practice. Although the final scheme did incorporate some elements of my proposal, albeit inappropriately applied, those inclusions offered no consolation whatsoever.

I continued to manage the project design through its execution, but have never had any desire to visit the site. Yet I have to admit that even though that result didn't turn out as I dreamed, the notion of being a genius did evoke a certain ongoing appeal.

A Mind-Expanding University Community

The University of Illinois Champaign-Urbana community exposed and engaged me in the company of exceptionally visionary, brilliant and interesting scientists and conceptualists.

Although there should be no doubt that marvelous minds and talents inhabit all university environments, I will theorize that those like the U of I which are located in relatively small self-contained urban settings might often tend to be more "communal" than those in larger metropolitan centers which disperse commuting resident-participants.

As I write, my mind is flooded with memories of many inspiring people who I honor as important formative influences during my early collegiate experiences. With sincere apologies to a great number of others, I will limit the following discussion to just three of them.

Lebbeus Woods

My College of Architecture classmate and closest friend, Lebbeus Woods, who I briefly mentioned earlier, was a radiant beacon of brilliance, originality and talent throughout his life. We were both financially-challenged married students at the time, each very serious about pursuing the intellectual and creative ideals of highest-possible level.

Leb was an authentic architectural studio superstar who must have won every top design award the college offered. His conceptual forms were often sparingly and strikingly simple— bold and powerful—even somewhat intimidating. His mastery of illustration—drawing and painting—left no doubt that he was marvelously committed, passionate and gifted.

Painting was Leb's greatest passion. I remember his apartment-stretched canvases, jars of casein paints and piles of empty Coca Cola cans which had previously caffeinated his sleepless obsessions to works-in-progress. They were wonderful, original expressions of a special world or worlds that he conjured in his mind. He gave me some of them, which I still own.

Leb never finished his undergraduate degree work. Instead, he packed up with his wife Patsy and a baby daughter and drove their old Buick to New York City. I'm not sure why he left. He didn't have any immediate work prospects there, and he certainly had no cash cushion. My best guess is that he had become bored with all-too-easy achievements of academic life.

Leb later told me that upon arrival in Manhattan he had virtually camped out near the offices of the prestigious Eero Saarinen architectural firm until they hired him. His first job with them was to become their field project manager for the Ford Foundation Headquarters that was undergoing construction during the mid-1960's. While there, he continued to paint, and also produced illustrative architectural project renderings for other firms.

We were reunited when Leb briefly returned to Champaign-Urbana to join me as a designer at Richardson, Severns, Scheeler & Associates. As discussed later, I deeply engaged with sculpture and industrial design at that time, and we had a two-person show at the University of Illinois campus gallery.

Leb soon returned to New York to concentrate his creativity on conceptual paintings and drawings. His single built project was a Light Pavillion created within a vast complex of towers in Chendgu, China by well-known architect Steven Holl.

Lebbeus Woods passed away in October 2012 at age 72. His loss as a creative force was eulogized by many others and his life and visions influenced in a tribute which appeared in The Guardian.[8]

As architect Nigel Coates reflects:

[Lebbeus Woods] reminded us that to believe

---

[8] *Lebbeus Woods, visionary architect of imaginary world, dies in New York*, The Guardian, Oliver Wainwright, October 31, 2012

in the existence of architecture, you need to feel it. Elaborate drawings of found spaces full of whirring sticks and lines of energy were genuine attempts to materialize the experience of space. Who else could do this? Nobody!

Architect and Yale Professor of Architecture Peter Eisenman said:

Lebbeus was one of the last of a generation of visionaries who dedicated a life in architecture to drawing an alternative world, one important for the present and the future. His singular mind and hand will be deeply missed.

Architect Zaha Hadid said of him:

His visionary work explored the fantastic potential and dynamism of space with radical proposals and powerful drawings that were extremely influential. His Light Pavilion in Chengdu will be a testament that our profession has lost a great voice.

Mark Morris, an architect and professor at Cornell University, eulogized Leb:

In words and deeds, Lebbeus reminded colleagues and students that drawing was the highest form and clearest expression of architecture. To watch his hand draft was to watch his mind at work. His lines were vectors, lines with direction and purpose, lines that danced. In an age of digitally

*powered representation, he powerfully
communicated with ink and graphite. Yet he
was never old school; no one took the future
more seriously than Lebbeus woods.*

Dr. Heinz von Foerster

As undergraduates, Leb and I were tremendously honored to be invited into a private circle of inordinately accomplished and diverse thinkers assembled by Dr. Heinz von Foerster, a legendary University of Illinois intellect. All other members, as I recall, were faculty from a variety of fields.

Heinz and his wife Mai hosted elegant and stimulating dinners for very small groups of mysteriously-selected scientists, artists, philosophers, poets and architects. A highlight was for each attendee to present a riddle of their own making to challenge and facilitate thoughtful discussion. These were wonderful occasions.

Dr. von Foerster, himself a noted physicist and philosopher, contributed great insights and clever wit to these conversations. His remarkably multi-faceted interdisciplinary ancestral and career background perfectly suited those unforgettable occasions.

Born in Vienna in 1911, his family had deep ties to Europe's intellectual culture of art, politics and science. Heinz's relatives included painter Erwing Lang, philosopher Ludwig Wittgenstein and playwright Hugo von Hoffman. His great-grandfather Ludwig Foerster had been a chief architect of the 5.3-kilometer-long Ringstrasse in Vienna, considered by many to be the most beautiful boulevard in the world.

While in high school, Heinz became engaged with a group of philosophers and scientists known as the Vienna Circle. He later studied physics at the Technische Hochschule in Vienna and at the University of Breslau, where he received a doctorate in 1944.

Heinz von Foerster's work laid the foundation for future research by others on a diverse range of sciences from biological physics to computer science—ultimately leading to his pursuit of understanding the nature of knowledge itself. In doing so, he formulated a set of philosophical ideas that later became known as constructivism.

In 1958, Heinz founded the Biological Computer Laboratory (BCL) at the University of Illinois as a leading international and interdisciplinary center for work in fields including biophysics, mathematical biology, computational technology, cognition and epistemology. There, his laboratory investigated a possible molecular basis for memory.

BCL also conducted pioneering work on parallel computing to break problems into multiple parts in order to speed computation. The first demonstration they developed, known as Numa-Rete, consisted of an array of photocells attached to a series of computer circuits that was capable of recognizing multiple objects.

In the 1970's Heinz contributed to more refined thinking in the field of cybernetics, the science of information theory. This work asserted that an observer can impact the system being observed—a great counterintuitive paradox of quantum theory.

As reported in the *New York Times*, Heinz von Foerster passed away at age 90 in 2002. A 5th International Heinz von Foerster Conference at the University of Vienna in October 2011 highlighted two technical sessions on his honor: *Self-Organization and Emergence in Nature and Society,* and *Emergent Quantum Mechanics,* citing him as a pioneer of this science.

Dr. William (Bill) Greenough

My life-long interest in cognition and memory is readily traceable to early influences of Heinz von Foerster along with another distinguished researcher who became a very close University of Illinois friend, the late Bill Greenough.

Whereas Heinz had introduced me to potential correlations between cybernetic and neurosciences, Bill's work explored biological and experiential aspects of brain plasticity and development as it relates to establishing the neural basis of learning and memory. I have discussed more about this phenomena, along with fascinating aspects of quantum theory studied by Heinz von Foerster, in my previous book *Thinking Whole*.

I met Bill soon after he had joined the U of I faculty in 1969 following his receipt of a doctorate in psychology at the University of California at Los Angeles. Upon arrival, he played a critical leadership role in the establishment of the multidisciplinary Beckman Institute for Science and Technology, served as one of its first two associate directors, and led its biological intelligence research theme for many years.

Much of Bill's research investigated effects of problem-solving stimulation, environmental enrichments, exercise, aging and injury upon the brain. These animal studies employed a variety of research tools and techniques ranging from optical and electron microscopes to electrophysiological and molecular approaches in order to understand how the brain responded to a variety of influences.

Bill's research earned him many high honors: He was named a fellow of the American Association for the Advancement of Science and of the American Academy of Arts and Sciences, and was elected to the National Academy of Sciences.

Dr. Greenough was also a recipient of the American Psychological Society's William James Award, the APA's Distinguished Scientific Contribution Award, the Society for Research Into Child Development's Distinguished Scientific Contribution Award, the Fragile X Foundation's William

Rosen Award for Outstanding Research and a National Institutes of Mental Health MERIT Award.

At the time he retired in 2009, Professor Greenough held a Swanlund Endowed Chair at the University of Illinois, was a Neuroscience Program director, and headed the Center for Advanced Study (CAS). Psychology Professor Neal J. Cohen, who succeeded him as Neuroscience Program director, said of him:

> *"Bill was one of the towering figures in neuroscience, not only on this campus but around the world," said Neal J. Cohen, a professor of psychology at Illinois and the director of the Neuroscience Program once led by Greenough. "His work led the way in illuminating experience-related plasticity in the mammalian brain, overcoming early views that sensory and motor systems of the brain were largely fixed very early in life, showing instead that the development of new synapses occurred in response to environmental enrichment and learning."*

Dr. Cohen's tribute to Bill added:

> *His research revealed that environment, exercise and training continued to shape the brain throughout the lifespan. The work led to new insights into the signaling and regulatory mechanisms at work in the brain and how those functions can go awry in conditions such as Fragile X syndrome, the most common cause of inherited mental impairment.*

Sadly and ironically, Professor Emeritus Bill Greenough, a true pioneer in studies of the brain, died in 2014 at age 69 of complications of dementia.

Why did I choose to remember and memorialize these individuals? What did I learn from their examples?

It's difficult to sum up their many lessons in just a few words, but I'll nevertheless share some brief thoughts.

Lebbeus Woods exemplified enormous power of self-determination—authentic, unlimited creativity—passion for ideas and ideals. He made me believe that Howard Roark was more than just a mythical spirit. He demonstrated that that natural genius within him might very well exist in all of us if we allow it to, and that if this includes me...maybe that's perfectly fine.

Heinz von Foerster epitomized and celebrated a whole-brain-embracing philosophy and life. He erased boundaries between art, science and philosophy, recognizing that science is also very much an art, and that philosophy (while subjective), offers broad guidance and meaningful context. He challenged boundaries between what is knowable and what is not—between the observer and the observed—between Newtonian logic and quantum theory.

Bill Greenough pushed me to recognize that our brains' neoplastic learning and memory capacities have a "use it or lose it" aspect that continues throughout our lives. Exercising our minds through constructive thoughts and behaviors reinforces physical and chemical synaptic connections along appropriate brain cell (neuron) pathways. This "what fires together, wires together" phenomenon occurs across our whole brain.

So go ahead—think a lot, do a lot—risk getting a big head.

Scene from Ayn Rand's Fountainhead movie

Lebbeus Woods

Dr. Heinz von Foerster

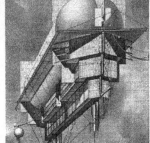

Lebbeus Woods: Marvelous visions of alternative worlds

University of Illinois Graduate School of Library Science concept

Proposed Eastern Illinois University dormitory concept

Chapter 3: Making Big Plans

WHILE AN UNDERGRADUATE student, two popular books of that period profoundly influenced me to expand my perceptual boundaries beyond prevalent studio views of "architecture" as defined by building structures or urban complexes to encompass entire metropolitan communities and regions.

Jane Jacobs' *Life and Death of Great American Cities* took issue with orthodox ideals regarding modern city planning such as Frederick Law Olmsted's-inspired City Beautiful Movement and Le Corbusier's Radiant City. These orthodox ideals emphasized architecture as individual monuments and skyscrapers assembled in park settings. Jacobs instead advocated an emphasis on more human-scale horizontal infrastructures with dense, diverse, vital mixes of public and private places—an interconnected environment where pedestrians can move about freely and unimpeded by automobile traffic and congestion.[9]

An even more important influence was the Jean Gottman's book published that same year titled *Megalopolis: The Urbanized Northeastern Seaboard of the United States.*

---

[9] *Life and Death of Great American Cities*, Jane Jacobs, First published by Random House (1961)

Gottman summarized a four-year study during the 1950's of the Northeast which he concluded:

> ...was the cradle of a new order in the organization of inhabited space.

The region was nearly 600 miles long, had an estimated 38 million inhabitants, and accounted for about one-tenth of the world's manufacturing and commercial activity.[10]

Megalopolis, which stretched from just north of Boston to just south of Washington, had gained many characteristics of a single city and had become what Gottman characterized as pre-eminent in the policies, arts, communications and economy of the United States.

The huge notion of high density urban areas comprised and visualized primarily as seamlessly interconnected pedestrian environments stretching over many miles held a powerful imaginative appeal. After all, I was a small-town guy who had come to believe that, despite density and geographic scale, cities might combine the humanistic vitality that Jacobs valued, with means to accommodate contiguous economic and cultural growth demands observed by Gottman.

Development of the Minneapolis Skyway System in the 1960's and 1970's offered a reference model that helped to overcome vehicle traffic barriers and boundaries which divide cities into building blocks comprised of captive boxes decorated by architects. The first of these were the brain child of real estate developer Leslie Park who, sensing pressure from emerging indoor shopping malls, wanted to create a similar environment in Downtown Minneapolis: a year-round climate-controlled space which enabled pedestrians to move

---

[10] *Megalopolis: The Urbanized Northeastern Seaboard of the United States*, Jean Gottman, The MIT Press, Cambridge, Massachusetts, 1961.

from building to building at second and third floor levels.

The system had grown by 1972 to link seven total segments, although many skyways remained disconnected from one another. The construction of the IDS Center in 1974 formed a "hub" featuring a spacious atrium area called the Crystal Court to unify the skyways in all four directions. The ultimate 11-mile-long network interlinked buildings over 80 full city blocks...the longest continuous climate-controlled pedestrian system in the world.

Synchroveyor Mass Transportation System

My pursuit of a self-imposed design challenge that produced Synchroveyor emanated from an observation that reliance on vehicular traffic arteries had caused cities to be broken up into nearly inaccessible islands of buildings which grope skyward in futile attempts to reach air that is free of the toxic fumes belched by cars, buses and trucks below. It followed that we could not hope to rehabilitate cities without first removing the cause of the deformities.

I wondered, might pedestrian transportation be integrally built into the very fabric of cities in much the same way that elevators are incorporated into individual buildings? And might such elevators be laterally oriented, instead of vertical, to avoid much necessity for high rise structures to accommodate densities?

Then, if so, could those integrated transport systems enable convenient, continuous and uninterrupted individual passenger transfers from all points everywhere along the entire network to all others? Could they safely accommodate very high-density demands more expeditiously than existing "moving sidewalks" provided at airports and shopping malls? Could they be integrated into both interior and exterior areas, be quiet, nonpolluting and unobtrusive?

Reframing these questions as design goals, I devoted a

great deal of time and thought over a two-year period to seek and develop a tangible solution. Throughout this period, I was also attending to full-time study schedule, plus part-time outside work demands.

The patented Synchroveyor mass transportation system concept which emerged from that self-imposed challenge may very well be my best technical idea ever. Although it is yet to be built, I expect that one day it will be.

The overall scheme features special configurations of articulated pallet loops that can be several blocks, or even miles, long. These loops, or "endless trains," are synchronized in particular movement sequences to enable passengers to be automatically transferred between them by means of rotating turntables. I produced mechanical designs for the system elements, along with a variety of illustrative Synchroveyor applications and settings.

I strongly believed that my Synchroveyor concept had merit, and contacted Howard W. Clement, the Chairman of the Board of Trustees for the entire University of Illinois System, who was a partner at a large Chicago intellectual property firm, to seek patent services. I explained to him that while I didn't have any money, I would owe him and pay in full when I could.

You may correctly observe that this was a rather audacious thing for an undergraduate student to do, but fortunately, he didn't see it that way. He personally responded within just a few days that necessary fees to pay for such services would be too high for me to even contemplate, but that his firm would donate those services at no cost.

Mr. Clement and his partners had been sufficiently impressed by the substance of materials I provided to select Synchroveyor as a project worthy of their support as a philanthropic contribution. He was true to his word. I prepared the patent drawings and Hume, Clement, Hume and Lee undertook the considerable legal work to secure a major patent.

Years later, my patented Synchroveyor received the 1970 national ALCOA Ventures in Design Award and was featured in many newspaper and magazine articles, including major architectural and engineering journals.

This prompted an invitation to visit my mentor, Howard Clement, at his top floor office in a Chicago high-rise building. Although we had exchanged correspondence on a few occasions, I had never met him before, or even spoken with him on the telephone. As a 4th year undergraduate who was meeting with a top university official whom I greatly appreciated and respected, this was indeed a very big deal.

I asked Mr. Clement why he had been so generous to me. He then walked to his window wall that overlooked the new Chicago Circle Campus of the University of Illinois and told me something I will always remember. He said he believed that everyone should have the opportunity to realize an important dream during their lifetime. That campus was his... it had been designed and built on his watch. He then said that he wanted me to realize my dream also, and that he hoped that I would do the same for others.

I like to think that I try to do this through teaching.

Crashing Through Mind Barriers

The Synchroveyor experience yielded early lessons and developments of immense personal significance. Among these, it taught me the great value inherent in mindful ideas is wasted if not acted upon, that the best rewards are hidden in challenges of pursuing worthwhile risks and that smart people risk asking dumb questions.

My original interest in the formative influences and connectivity of transportation within urban architectures was clearly not shared or encouraged by faculty advisors who failed to see any professional relevance. This being the case, my work leading to Synchroveyor was considered to be entirely outside mainstream boundaries of legitimate architectural design. Consequently, I neither expected nor received any academic credit for this outside activity.

At the same time, the ultimately patented Synchroveyor concept evidenced some very sophisticated mechanical attributes which no prior training had prepared me for. Apart from replacing the brake pads on my old cars a few times (yes, before disc brakes replaced them) that was about it.

Recognizing that I needed help with certain key technical elements, I also prudently realized that gaining informed assistance would require that I be able to conceptually illustrate what I had in mind as clearly as possible. This involved risking exposure of immature thinking—potentially dumb ideas—in order to receive meaningful outside inputs.

Most fortunately, I picked a perfect person to present the resulting voluminous conceptual detail sketches and descriptions to—a senior professor in the University of Illinois Mechanical Engineering College. The wonderful advice that Dr. Wayne Adkins offered was thoughtful and illuminating.

Upon reviewing my proposals, Dr. Adkins told me that my mechanical solutions made good sense, because engineering is all about common sense. He advised that I should never rely upon others to do fundamental thinking as I was fully capable of accomplishing by myself, but also to have the wisdom to seek ideas and feedback from others with more specialized experiences and skills. He generously offered to provide those conceptual refinement services provided that I continue to do the conceptual heavy-lifting.

And that's exactly what followed as I later developed numerous highly-detailed patent drawings.

My Synchroveyor design pursuit led me on another marvelous learning adventure involving the University of Illinois Physics Department. Here, however, I'll briefly diverge from that technically-correlated discussion to share a quite different personal physics lesson which wasn't covered in my undergraduate class textbook.

I had developed a rather keen interest in physics during my early undergraduate years which included some inherent philosophical aspects. Believing that I had discovered a wondrous but since forgotten metaphysical implication of 19th century English physicist Faraday's Law of Induction, which describes how electric current produces a magnetic field and vice versa, I was eager to explore the ramifications more fully with someone with great expertise who might share my excitement.

I managed to arrange an appointment with a generously accommodating University of Illinois Nobel laureate professor of physics to seek his wisdom on the matter. He listened tolerantly to my seemingly profound theory with only slight evidence of amusement—and then offered a very honest response I hadn't expected.

As now clearly recalled, he said: "Larry, although that's interesting, I really never think of such matters. I chose to pursue my research in physics simply because I enjoy solving scientifically-bounded puzzles with definitive proofs."

That deflating lesson taught me that not all physicists were seeking divine spiritual revelations as I had naively imagined after all. There was no Howard Roark to be found there.

Back to Synchroveyor...dissatisfied with a highly mechanical solution I had originally conceived to transfer power to mobilize the system, it occurred to me that this might be accomplished far more simply ("elegantly") through electromagnetic propulsion and levitation.

While taking a basic undergraduate-level physics class, I began thinking about an idea of re-visualizing a conventional electric motor which consists of an armature rotating in a magnetic field to instead imagine an armature pulled laterally along an adjacent magnetic field. Such a linear field, if incorporated along the Synchroveyor support structures with armatures attached to the moving system, could accomplish exactly what I wanted with no moving parts or supporting wheels whatsoever.

I discussed the revolutionary idea with a physics class instructor who became very enthusiastic. As it turned out, however, there was one significant reason for disappointment. The linear induction motor had already been invented—it was first used in 1969 on the Japan's Chuo Shinkansen MagLev bullet train.

Nevertheless, there was a personal happy ending to another part of this story—one which had begun badly.

I invariably tended to score poorly on physics exams under time pressures which required rote categorical memorizations of formulas rather than visualizations of appropriate problem-solving principles. These circumstances caused me to mentally choke up, leaving me without adequate time to think through all of the test questions.

Soon after my final exam, I received a surprising telephone call from my instructor informing me that he had scheduled a meeting for me with the head of the physics department later that same day. When I asked why, he replied that I would find out during that meeting.

As it turned out, and as I had fully suspected, the purpose of that appointment was *not* to congratulate me on my final and overall exam results. My class scores had ranked close to the bottom of a huge student population.

When the department head then informed me that he had decided to pass me despite my lousy grades, I foolishly questioned why he would make such an exception. He answered that it was a shared decision including the urgings of two senior faculty who had come in contact with me. Both had independently contacted him, each unaware of the other, saying "we can't in good faith flunk this guy."

The department head then closed the meeting saying that

his personal reason for the decision was because he was leaving on a family vacation the next day, and he wanted to start off in a good mood.

So maybe some physicists, Faraday included, have a strong spiritual streak after all.

Revitalizing Decayed Urban Cores

John Steinbeck, in his 1960 book *Travels with Charley: In Search of America,* made an observation regarding the decay of cities which echoed the observations of Jane Jacobs:

> *When a city begins to grow and spread outward from the edges, the center which was once its glory...goes into a period of desolation inhabited at night by the vague ruins of men. The lotus eaters who struggle daily toward unconsciousness by the way of raw alcohol. Nearly every city I know has such a dying mother of violence and despair where at night the brightness of the street lamps is sucked away and policemen walk in pairs. And then one day perhaps the city returns and rips out the sore and builds a monument to its past.*

This same decay and renewal theme was addressed earlier by Le Corbusier:

> *The center must be modified in and out about itself. It crumbles and rises up again through the ages; just as a man changes his skin each seven years and the tree its leaves year by year. We must concentrate on the center of the city and change it, which is after all the*

simplest solution, and more simply still, the only solution.[11]

Austrian-born architect Victor David Gruen, who was noted as a pioneer in urban revitalization proposals, placed key importance on incorporating transportation systems which were designed to serve specialized needs of central cities. He said:

> *A sign of good health in a large core area is a transportation system specifically adjusted to the needs of a compact area of great vitality...rather than a carbon copy, in reduced scale, of transportation systems designed to achieve easy accessibility to and from the outside.*[12]

Gruen continued:

> *If we are to succeed in the creation of a new, diversified urban pattern, we will require a new and diversified means of public transportation. As we create public areas with different characteristics and new types and arrangements of urban elements, we will need transportation types scaled and engineered to the specific uses for which they are destined.*[13]

---

[11] *The City of Tomorrow and Its Planning*, Le Corbusier, Payson and Clarke. Ltd., New York,1929, p. 47.

[12] *Heart of Our Cities*, Victor Gruen, New York: Simon and Schuster, 1964, p. 85.

[13] *Heart of Our Cities*, Victor Gruen, New York: Simon and Schuster, 1964, p. 245.

As I too recognized at that time, mass transportation can be of limited benefit unless systems are designed specifically for the demands of high density areas. Buses and trains, including monorails, are designed for moving people over relatively long distances, and their efficiency increases with the distance between stops. Essentially, they are no more than elongated automobiles.

In 1964, prior to developing the Synchroveyor concept, I had begun to wonder if a system of mass transportation could be "built-into" the existing urban pattern in a way which would eliminate much of the congestion and provide a circulation system upon which a healthier urban organism could grow. I intentionally limited my objectives to the movement of people, rather than "general transportation" out of a conviction that cure-all solutions often cause more problems than they solve.

I assigned myself a problem statement titled *Design an Appropriate Primary Distribution System for People in High-Density Urban Areas.*

After first asking myself: "What characteristics should an ideal system have?" I then defined the following objectives:

1) Convenience:

- The system should provide maximum flexibility in allowing passengers to travel to individual destinations without requiring time-consuming and inconvenient en route transfers.

- It should be readily accessible for short distance intervals.

- Boarding opportunities should be frequent and completely dependable.

- It should be easy for all people to adjust to, including physically handicapped people,

elderly people and small children
- The system should enable passengers to arrive close to their destinations.
- If the system is to be "built-into" a city, it should be a distribution system for moving people short distances, and also provide for express passengers who should not be subjected to numerous intermediate stops.
- It should be accessible from within city buildings.

2) Capacity:
- The system should accommodate large numbers of people without contributing to congestion.
- It should be able to adjust to varying load demands which occur during different periods of the day.

3) Safety:
- The system should be safe for all persons. It should remove dangers which could result from errors in judgment or human disabilities.
- The proposal should consider the protection of passengers from social violence which has been associated with mass transportation in metropolitan areas.

4) Comfort:
- The system should provide for passenger comfort. It should operate in a climate-controlled environment. Seats should be provided, particularly for express passengers.
- Smooth mechanical operation will contribute to

passenger comfort by eliminating lurches and vibrations.

5) Dependability:

- The system should be mechanically dependable. The design must consider vandalism as well as normal wear.

- In-service repair and maintenance should be provided for on primary mechanical and power systems so that a partial breakdown will not disrupt operation. Back-up power should be provided.

6) Flexibility:

- The system should be adaptable to existing cities. It should enable implementation in stages and should be expandable.

- The system should be able to change elevation to operate underground, or at any level.

7) Speed:

- The speed of the system should be expeditious but must allow passengers adequate time for orientation to destination locations.

8) Environmental Desirability:

- The system should be clean and quiet. It should not contribute to air pollution or to noise levels in cities.

- The system should not aesthetically violate the area through which it passes.

9) Practicality:

- Ideally, the system should be inexpensive to

operate. The cost for the system must be compared with the economic productivity which would be expected to result from its implementation in terms of value considerations.

- Individual passenger fares should be low to encourage its use.

10) Integration with Other Transportation Modes:

- No single solution can be expected to provide for the total transportation needs of a metropolitan area. The only successful approach must be one which integrates specialized systems into a larger diversified program dealing with all forms of transportation. In compliance with this objective, the system can tie together commuter rail stations, bus stations, and all other transportation facilities within the central area of a city.

As I wrote at the time:

While architects and planners are generally credited with development of urban spaces, traffic engineers have consistently had the greatest roles in determining the forms of cities. Perhaps this is only right, for in cities, Automobile reigns supreme.

Subservient man's attempts to appease Automobiles' insatiable greed for space have been spectacularly unsuccessful. Streets and parking facilities are such heavy consumers of land in city centers that if enough space is to be provided to meet ever-growing demands of

*traffic, there will soon be little reason for man
to visit these areas. We will find only parking
meters, projected from oceans of asphalt.*

*It is in the city centers, amid the few
remaining architectural glories of the past
which have not yet been converted to parking
lots, that Automobile has most clearly
demonstrated its superiority over man.*

Quoting British Minister of Transport Ernest Maples, I added:

*When a ton of steel moving at 30 miles an
hour and eleven stone [154 pounds] of flesh
and bones moving at three miles an hour share
the same surface, accidents must happen, and
the flesh and bones never win.*[14]

Revolutionary Capabilities and Capacities

The Synchroveyor mass transportation system represents a major departure from conventional vehicular and moving sidewalk solutions. I'll briefly explain why this is both true and beneficial.

First, in order to visualize the concept's most significant and unique characteristic, imagine we are looking at an aerial plan view of a movable mechanical loop (or "endless train") which operates at speeds which range from standstill to any appropriate maximum speed—let's say 20 miles per hour.

In order to provide an idea of scale let's assume that the length of each roughly square or rectangular loop is a distance of three city blocks (approximately 1,200 feet) and that the

[14] *Heart of Our Cities*, Victor Gruen, New York: Simon and Schuster, p. 116.

mechanical loop is six feet wide. While we're at it, let's also assume that during each moving sequence, any point in that continuous loop will advance a travel distance of three blocks. This means that any person located within an area encompassed by the loop would be within 600 feet (half of that particular loop length).

Since the mechanical loop is continuous, passengers can board from anywhere along it when it is not moving.

Now we'll take the concept to a multi-loop level where one set of loops—the local ones that stop and start—are synchronized with adjacent constantly-moving express loops encompassing multiples of those local loops so that passengers can transfer back and forth between the local and express sets when they are moving at the same speeds. Passengers can also transfer back and forth between express loops via local loops.

Working all these schematic synchronicities for a variety of different plan applications and configurations without computer simulation was a mental brain-twister which I laboriously calculated using numerous time-sequenced drawings. This proved to be a much more complex challenge than any posed by mechanical design demands.

Altogether, this defines the general schematic Synchroveyor concept, with one additional key design feature. Passenger boardings and transfers would occur simultaneously and automatically via closely spaced rotating transfer decks applying a "lazy Susan" approach along their lengths. Low partitions would prevent riders from entering/exiting at the wrong times.

As for capacity, let's now assume that local transfers would occur each minute, so that theoretically, an estimated 241,200 people can board a six-foot-wide accelerating loop—and exit—within each mile of the system during each one-hour period. Back at the time I conceived the idea, that number exceeded the population of Lansing, Michigan.

Comparing Synchroveyor with other transportation

modes from an urban space-utilization perspective, let's first estimate the area needed to accommodate one million people per day relying exclusively upon private vehicles. Assuming that the drivers and passengers would occupy about 750,000 automobiles, parking alone would require approximately 150,000,000 square feet of space.

If that city was Manhattan, all structures in the main business district would have to be demolished—plus a large area of Manhattan Island would have to be double-decked, just to provide necessary parking spaces.

And whereas a fat man standing comfortably on a subway requires only about five square feet, that same portly fellow— or a skinny one—sitting in a parked car requires 200 square feet. If the car was moving at 30 miles per hour, the space it would require is approximately 600 square feet (allowing one car length per 10 MPH), and 1,200 square feet at 60 MPH.

Now comparing the amount of circulation space required to move 7,000 passengers along a one-mile route with cars or buses versus a single six-foot-wide accelerating Synchroveyor loop:

- This would require 34 vehicle lanes of cars (1.5 passengers per car) spaced at 20 feet.

- It would necessitate four lanes of buses spaced at 20 feet.

- Synchroveyor could accommodate all in one-half of vehicle lane.

Some Illustrative Urban Applications

In the summer of 1966, I presented a proposal describing a Synchroveyor installation for the Chicago "Loop" area to various transportation and planning officials in response to a need expressed in a "Chicago area Transportation Study

Larry Bell

(Volumes I, II and III.)." As expressed by the City of Chicago:

It has been proposed that, in the Central Area, the distance from railroad station to desk be significantly reduced. This could be accomplished by means of a belt or some type of continuously moving system serving as a high-capacity horizontal elevator. Such a system would not only bring the railroad stations closer to the Loop buildings, but would, in effect, bring the buildings closer to one another. Improved travel time within the Central Area has the potential of reducing the Area's time size or of enlarging its physical size.

The study went on to say:

These pedestrian ways are proposed either above or below street level. They will, thereby, withdraw much of the pedestrian action from the surface streets. This has the virtue of separating pedestrian and vehicular traffic by putting the lighter, more easily-manages traffic (the pedestrian) on raised structures or into tunnels.

The report continued:

The virtue of such a system lies in its ability to cut travel time and to increase personal safety and pleasure. The system is laid out to connect with commuter rail stations and thus would be geared to aid those persons who must walk the greatest distances from rail terminals to Loop buildings.

This sounded to me like a perfect recipe for Synchroveyor.

My response proposed two local Synchroveyor loops interconnected by a single express loop incorporated into all-weather pedestrian ways. I suggested that the passageways crossing Michigan Avenue along Jackson and Randolph streets be extended to Canal Street, and also that another pedestrian way be built along Canal between Jackson and Washington.

My plan for Chicago also proposed the development of special structures (essential surface-level tunnels) to enclose vehicular traffic and isolate the noise and exhaust fumes from newly-created pedestrian spaces. These were envisioned as a temporary solution to the problem of banishing vehicular traffic...the ultimate solution being to provide more desirable specialized means for distributing goods within the city. (As later discussed, a few years afterwards I headed a study which proposed doing exactly that within an existing network of underground tunnels.)

But there was a major problem with all of this... Synchroveyor didn't exist as an immediate reality. It existed only in the form of Patent No. 3,421,450.

Other proposals were developed to illustrate Synchroveyor applications for other smaller environments. These included three alternative single-loop Synchroveyor plans to rebuild and revitalize a core area of Jacksonville, Illinois as a hub of commercial, civic, cultural, and recreational activities.

Another single-loop adaptation was proposed for the Boston Logan international Airport, and a multiple-loop scheme was prepared for the Dallas-Fort Worth Regional Airport.

Still another was proposed and considered by the University of Illinois Champaign-Urbana campus, which brings my life by design story into a consequential new chapter.

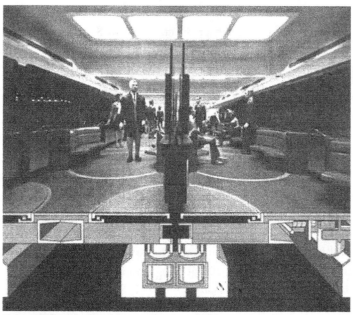

Patented Synchroveyor mass transportation system

Automated passenger transfer   **National Alcoa Award, 1970**

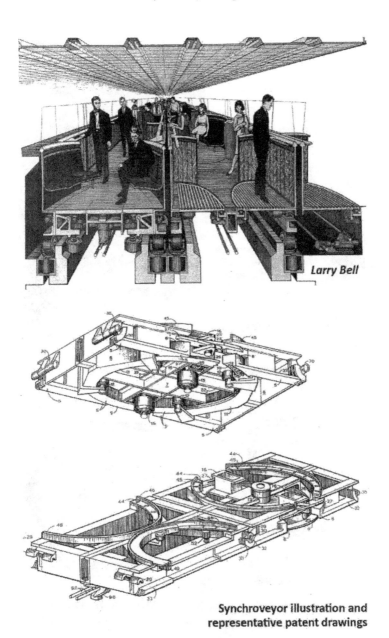

Larry Bell

Synchroveyor illustration and
representative patent drawings

Larry Bell

Synchroveyor: Chicago Loop application

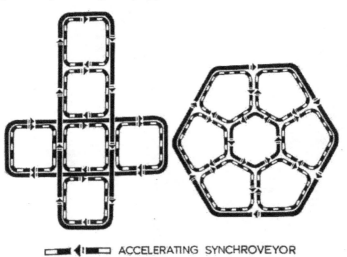

ACCELERATING SYNCHROVEYOR
CONST. SPEED SYNCHROVEYOR

Larry Bell

Synchroveyor: Chicago Loop application

Synchroveyor: Chicago Loop application

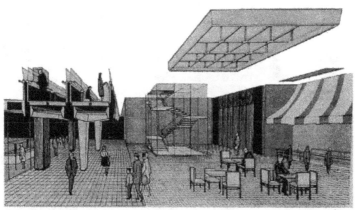

Synchroveyor: Interior building view

Larry Bell

Larry Bell

Synchroveyor: Interior building view

Larry Bell

Synchroveyor: Street with enclosed traffic

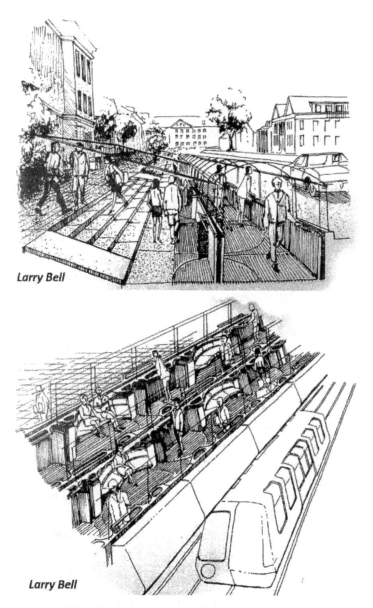

Larry Bell

Larry Bell

Proposed Synchroveyor for U. of Illinois Champaign-Urbana Campus

Chapter 4: My Transport To Industrial Design

AS I REMAINED at the University of Illinois expecting that a proposed Synchroveyor would be implemented on campus, that stay turned out to be a great deal longer than I had imagined, and included some enormously consequential personal and professional life changes.

A recent divorce with Sally had dropped me into a very dark period of emotionally comatose depression. There was no blame or acrimony on either side—there never had been—and although geographically distant, we retained a caring friendship throughout her life.

The U of I Department of Urban Planning had been actively recruiting me to enroll as a graduate student, and I briefly accepted that invitation. I left the program a few months later upon determining that planning processes didn't offer the immediacy nor control of project results that my restless impatience and desire for creative independence demanded.

My next decision, albeit one guided by circumstantial convenience, set my life by design off on a marvelously fresh and satisfying new trajectory. While essentially biding my time in Champaign-Urbana, still working for Richardson, Severns and Scheeler, I enrolled in the U of I Industrial Design Graduate Program.

Although the Industrial Design program was located on the same campus as the Architecture and Urban Planning program, the transition to the College of Fine and Applied Arts opened up an entirely new cultural and performing world. It was also exactly what I needed to raise me out of deep doldrums and non-constructive self-pity.

Quite in contrast with the more ideologically pompous atmosphere of my previous architecture classes, and the plodding bureaucratic conformity which I perhaps unjustly feared in the planning curricula, industrial design was a new wonderland of innovative tinkering, joyous creativity and even some good-natured mischief.

The late Edward ("Ed") Zagorski, the industrial design program head, was a charismatic fellow with eternal child-like curiosity about virtually everything and a delightful love of good fun, set the atmosphere. As Craig Vetter, one of my fellow students, recently recalled of Ed:

> *Here was a child...in a man's body. In his eyes, everything had some kind of wonderful possibility beyond what was there. One day: 'Here is a piece of wood to [help you] feel good.' Another day he handed us some wire and little squares of glass and said: 'Make a structure.' Another day he brought in some weather balloons, and said 'Let's stretch this rubber over structures and put fiberglass on it...'*

Craig Vetter, a grad school classmate, became a well-known motorcycle designer who was inducted into the American Motorcycle Association Hall of Fame. In 1998, his design for the British Triumph Hurricane was selected to tour the world in the Guggenheim Museum's *The Art of the Motorcycle* exhibit.

During the early '70s, while still in grad school, Craig designed and developed a now-ubiquitous combination motorcycle windbreaker-wind shield famously known as the Vetter "Windjammer" fairing. Similar devices later became standard equipment offered by all motorcycle producers. By the time he sold his company, the Vetter Corporation had become the second largest motorcycle-oriented manufacturing company in the United States. Only Harley-Davidson was bigger.

Ed's most famous undergraduate class assignment came to be featured in the April 12, 1963 issue of *Life Magazine*. The assignment was inspired by John Glenn's historic orbital Mercury flight, the first by an American, on February 20, 1962. The students were challenged to devise the means to land raw eggs unbroken that were launched high in the air by a catapult constructed by Craig Vetter from the leaf suspension spring of an old Pontiac. Targeted splashdown was in a reflecting pool "ocean" in front of the Krannert Art Museum.

The convivial student-faculty environment was a natural breeding ground for mirthful pranks. On one occasion, a professor was horrified to look out the window to see a large dent on the side of his beloved model 356 Porsche. The ghastly blemish had been convincingly painted with a washable pigment by one or more of the department's many very highly skilled illustrators.

Sadly, I can't claim any credit.

Parallel personal developments at that time changed my life very happily and dramatically. I married Cindy and gained a totally wonderful freckle-faced ten-year-old daughter, Dianne, in the bargain.

While Cindy was pursuing a doctoral degree in Social Work, Dianne ("Dee") became a star high school cross country runner who won more than foot races. Together with our Champaign, Illinois Unit 4 School District co-plaintiff, our family won a lawsuit against the Illinois High School

Association which resulted in a landmark ruling regarding discrimination against women athletes which changed policies throughout the United States.

I first became intensely engaged with sculpture during that eventful period. In addition, I secured a full-time appointment to develop a two-year art and design program at a Parkland College, a new Champaign, Illinois community campus.

The fact that we operated out of rather makeshift mixed-disciplined facilities prior to the construction of impressive campus structures served a very serendipitous advantage. For example, next-door nursing teachers taught anatomy for my drawing courses, the dental lab technicians taught our students how to cast metals, and the creative spirit of art and design flourished everywhere.

Several of my Parkland students later completed undergraduate studies in industrial design at the University of Illinois. One later joined the faculty.

In contrast with personal art, industrial design imposed a need to address a broad variety externally-imposed requirements. These considerations can include gaining an understanding of: who the intended product users will be; allowable pricing vs. costs of manufacture; numbers and means of production; safety, liability and human factor issues; competition in the marketplace; and distribution networks.

In some instances, these requirements may be modest, allowing the designer a lot of discretionary latitude. In other cases, the requirements can be very challenging, exacting and critical, and must be carefully heeded to avoid disastrous consequences. It is essential to understand unique conditions presented by each special circumstance.

My activities as a grad student developed some furniture designs I fabricated using thin walnut strips that were laminated into bent forms. The parts were then assembled together and hand carved into finished shapes. This process was quite complicated and time-consuming, and while personally satisfying, really didn't lend itself well to mass production at moderate costs.

Later, after graduation, I established a very small consulting organization, the Illinois Design Collaborative, that consisted primarily of myself, and began to design products that could be more readily and economically mass produced. Included were a number of concepts and prototypes for David Edward, Ltd., a manufacturer of residential furniture located in the Chicago area.

I also designed some fiberglass rocking chairs and created forms for molds using the same wood laminating and carving methods I had developed for sculpture I was doing at the time. I then paid to have them produced in limited numbers by a local boat company for sale to some high-end retailers. This was my first commercial business venture, albeit not a very profitable one. The item fabrication costs were still quite high due to a labor-intensive fiberglass hand lay-up method we employed. And the investment required to set up much less expensive injection molding capabilities were prohibitive for me.

Furniture is generally a very price-sensitive type of product, making cost control very important. Domestic furniture presents special challenges due to limited sales demands that limit mass production opportunities which can improve economies. Commercial furniture can be much more profitable because purchasers tend to buy it in large lots that enable the producer to operate on smaller unit revenue margins.

Form appeal, comfort and functionality are always important. But it is also essential to design products that respond to special interests and needs of sufficiently large target markets and which are reasonably simple to manufacture within cost and process constraints.

Most of the furniture concepts I developed for David

Edward, Ltd. were designed for relatively simple wood fabrication. Some pieces had plastic laminated side panels featuring easy pin-together assembly for efficient packaging and shipping.

Many of the chairs were applicable for either residential or business markets, including seating for company reception areas.

Desks, credenzas and tables were constructed of molded plastic shells reinforced with wood or fiberboard internal cores. The elements could be reconfigured or put together in different modular combinations for versatility and cost control. Most were intended for home settings and contemporary reception areas.

Traveling a New Zen Trail

My graduate student industrial design years provided grateful opportunities to explore and pursue a broad range of new interests. Most important among these, I discovered my innate passion for art. I developed a lasting love of hand-crafting forms from wood and other natural materials, and I stumbled upon a transformative book titled *Zen and the Art of Motorcycle Maintenance*.

Written in 1974 by the late Robert Pirsig, *Zen* tells a fictitious account of a real 1968 motorcycle trip Pirsig took from Minneapolis through the West with his 11-year-old son and two friends. A fifth unseen traveler named Phaedrus, who is actually Pirsig's alter ego, is brilliant, uncompromising and obsessed with searching for truth.

Subtitled *"an Inquiry into Values,"* Pirsig's novel isn't really so much about Zen or motorcycles. It's about living a good and meaningful life...about seeking possibilities to unify the cold, rational world of science and technology with the human and intuitive realm of art and spirit. It's about bridging between reason and emotion, between objective and subjective and between classical and romantic.

And it's an ode to sensing the world viscerally as experienced on a motorcycle, compared with the TV-like passivity of looking out the window of a car.

So, I bought a motorcycle.

The 400cc Suzuki wasn't big and powerful by Hell's Angels standards, but it was plenty fast for me. That lean and beautiful mechanical steed became an extension of my mind and body as it propelled me forward in great surges of energy.

With almost no thought or effort on my part, it tilted its angle of attack on curves as I unconsciously shifted my balance, and it hurtled me through wind and space. Or, when I wished, I could hum along at a leisurely pace and absorb the sights, sounds and aromas of my surroundings.

Unlike cars, which enclose and isolate you from the outside world, motorcycles put the world right in your face.

The sky revealed its fuller character, constantly being transformed by a changing display of cloud formations, colors and light conditions. Rain had a richer aroma out there also, often mixed with the somewhat acrid smell of ozone and heavy scent of wet humus. And those raindrops didn't just fall; experienced on the motorcycle, they became projectiles that stung exposed skin with ballistic force that punctuated their individual importance.

The land was constantly changing as well. Blankets of snow and ice dissolved to reveal rich black topsoil, purple rocks, and scattered patches of vegetation that had survived hard freezes. Sounds of tractors became audible over the steady putter of the Suzuki, and farmers dutifully waved back to me along country roads as they cut furrows for another planting season.

Fields and trees budded into blue, yellow and brown-tinted shades of green that gradually covered the Earth and contributed sweet fragrances, finally becoming dominant forms on the surface and horizon. Cornfields eventually reached heights where they obstructed views of road intersections and traffic, requiring that my riding become more vigilant.

I felt contented and free, and I was moved by the fertile richness of those Illinois plains.

Country life along those small asphalt and gravel roads added a special human dimension to the experiences. People appeared to be going about their affairs in the same general fashion that has characterized life there for generations. Farms expressed family cooperation, stability and pride that warranted admiration.

Old homes and businesses in various states of repair reflected histories of dreams, plans, lives and relationships that stimulated imagination. Residents in roadside towns where I stopped for coffee pursued timeless conversations and repartee which will continue so long as neighboring friends meet. I had no desire to interrupt them with my presence, but I was happy to be among them.

When I later moved to Houston traffic and gave up that motorcycle, I felt like I was abandoning part of my spirit. It had transported me to wonderful places, experiences and realizations that could never be repeated in quite the same way without it.

I would like to believe that much of that old motorcycle spirit part of me is still there. I am occasionally reminded of this when I walk or drive in the country or eavesdrop on a conversation in a small-town grill en route to somewhere else.

But for that time being, my life by design motorcycle journey remained based in Champaign-Urbana.

From Grad to Program Head

Immediately following the completion of my industrial design graduate studies, I was offered a position to head up that same program at the University of Illinois. This was an exception to university policies, as the university did not condone hiring its own graduates. In my case it was even worse because that was my second U of I degree...the first being in architecture.

The Assistant Professor rank they offered was also problematic because it required a two-year minimum appointment that exceeded the term period established by the faculty search process. Accordingly, I was asked to sign a resignation letter that would become effective after one year which the Dean kept in his desk. Apparently, he must have forgotten to open up that particular drawer during the next seven years during which I was promoted through the ranks to tenured professor status.

The industrial design graduate program emphasized problem-solving projects that addressed a wide variety of themes. Some dealt with concepts and approaches for creating new types of architectural structures, including modular housing components. Others addressed technological systems and applications, such as an ocean water monitoring device to support ecological research.

Many projects were aimed at offering better alternatives for elderly and other physically impaired people whose special needs were not being adequately served by existing products. Design responses included a powered wheel chair that can climb over curbs and other obstacles, and another wheelchair that can elevate the user to a standing position for accessibility to kitchen, bathroom and work counters. Another project developed and demonstrated a road vehicle that can be accessed and operated by a person without use of arms or legs.

Mass Transit System Studies

Several of my industrial design graduate students at the University of Illinois undertook ambitious transportation design and planning projects that hark back to my previous interests in architectures of movement.

Some produced concepts for vehicles that can be automatically dispatched to destinations on special guide way structures, or alternatively, can be operated under manual control. These studies were invariably quite comprehensive, addressing vehicle construction; interior layout, safety and comfort features; and vectoring control mechanisms.

A very major study involved a joint collaboration of graduate industrial design and architecture students in the planning and design of an automated transit system for people and cargo within the previously-mentioned Chicago freight tunnel network. This "Chicago Tunnel Study" was undertaken in cooperation with the Department of Public Works.

Although the resulting plan was never realized, it did receive a substantial amount of media attention after the Department's Director publicly announced strong interest during a news interview.

Another team effort was undertaken for the City of Minneapolis to design and integrate an elevated people mover system into the downtown skyway system infrastructure.

The Dragonfly

While teaching at the University of Illinois, two friends and I started a small company to build a "Dragonfly" car I designed.

Steve Norcross was actively involved in designing and building Formula One racing cars along with a variety of other products. Most were masterfully molded from fiberglass, including all of the fiberglass furniture I had designed.

Dr. Morton Tabin, a psychiatrist and my all-time most lovably eccentric pal, was Steve's racing partner and driver. Mort's exceptional race track success record was likely attributable to his unflappable lack of concern about competitors who were hell-bent to overtake him. Even more, Mort was unflappably unconcerned about almost everything—cars being a primary exception.

Mort loved cars to the point of physical peril.

Once, as we walked together, he became so transfixed by the sight of a parked unlocked car that he just couldn't resist getting in. When the owner arrived to find a total stranger sitting in his beauty I fully expected to witness a very ugly scene.

Yet Mort Tabin was not, after all, an ordinary stranger. What I instead witnessed was a fellow beaming with appreciation and pride as he and my enthusiastic friend discussed the car's admirable features.

Incidentally, having mentioned that Mort was eccentric as well as a total car nut, he was strangely obsessed with Volvos. He owned several of them. The affection he held for his BMW motorcycle seemed far more understandable, and he was a great biking companion on those Illinois prairie roads.

Mort Tabin lived and expressed irresistibly delightful child-like enthusiasm and innocence. I recall occasions when he would walk up to a table of unknown restaurant diners, inquire if they were pleased with the food selections they had ordered, ask if they would mind if he sampled some from their plates and I'd watch them cheerfully oblige.

Mort told me that he enjoyed providing voluntary psychiatric counseling with a local prison mental ward because the inmates trusted him. He and they had a natural rapport because they understood each other.

Now that I have already digressed to affectionately memorialize this beloved friend, I'll wrap up this nostalgic diversion with a final story which was recounted to me by his wife Kay.

Mort had collapsed on the floor of their home on the day before a brain tumor tragically ended his life. Unable to lift him onto a bed, Kay put a pillow under his head and asked, "Mort, are you doing alright?"

He answered, "Ahhh...I make a living."

That was pure Mort Tabin...right to the end.

Okay...back to Dragonfly. The car was totally designed and engineered from the road up to demonstrate that inexpensive, energy-efficient automobiles can be versatile, safe and enormously fun to drive.

We demonstrated that our car could be built at that time within a cost of under $4,000.

Dragonfly exhibited numerous novel design features. By simply sliding or removing its two-piece canopy, the car was quickly and easily transformed from a hatchback sedan to a touring landau, sports convertible, open roadster or utility pickup.

A specially-designed seat automatically slid up to enable users to step into the low vehicle without the need of heavy and complicated doors. They then slid down into the operational position under the user's body weight.

The spare tire served as the front bumper, and along with a unique body structure, afforded good crash protection. Rather than using a conventional steel chassis, the vehicle had a double shell (monocoque) fiberglass body with a plastic foam filled inner cavity. This resulted in a very light and strong structure weighing about 1,250lbs that yielded 40mpg efficiency. Dragonfly's four-cycle, 93 horsepower engine, placed at the center of gravity, provided excellent cornering performance with plenty of zip.

A sturdy roll bar was incorporated, and since entry was accomplished through the hatch, there were no doors to fly open in a crash. Motorcycle-type fenders mounted to the front wheels were designed to shear off in the event of a glancing forward collision as an additional passenger safety precaution.

The fully-independent suspension system was the same type used in the most advanced Grand Prix racing cars, with all parts specially manufactured or modified to meet Dragonfly specifications. Fully independent rack and pinion steering in combination with all four-wheel disc brakes provided quick and positive driver response.

The first Dragonfly we built demonstrated that its design and engineering features performed just as planned. Unfortunately, the first one was also the last. Our Mayview, Illinois fabrication facility burned to the ground along with all body molds, some engines, and other parts soon after the prototype was completed. This put a premature and unhappy end to the venture.

Getting Down to Business

The Dragonfly experience, along with business and legal lessons derived from other product design and patenting ventures and adventures, channeled my hard-wired inner Capitalist proclivities. These influences motivated me to begin writing a regular *Getting Down to Business* column as a Contributing Editor for Industrial Design Magazine, the primary U.S. publication in that field.

Writing incentivizes me to seek out information about topics I wish to know more about. In doing so, it also challenges me to become more aware of relevant but not broadly recognized connections with other subjects and ideas, and to advance my understanding and organize thinking about important lessons and implications in order to meaningfully share them with others.

Writing has become a life-long habit...something I obviously do a great deal of.

I continue to believe that mature professional designers have the requisite resources to be true leaders in any business environment, whether working in a company, a private practice or pursuing entrepreneurial ventures. These valuable attributes include big-picture vision, aesthetic sensibilities, multi-media communication skills and proactive demeanors.

Although personally eschewing time-demanding organizational commitments, those structured around painfully endless committee meeting cultures in particular, I have endeavored to support high qualitative and ethical standards of practice through membership in several professional societies. While heading the University of Illinois Industrial Design Graduate program, I was also elected to serve on the national board of the Industrial Designers Society of America, its executive committee and as a national IDSA design competition jurist.

Larry Bell

Street entry model

U. of Illinois automated transit proposal for Minneapolis, Minnesota

U. of Illinois automated transit vehicle design for Chicago tunnels

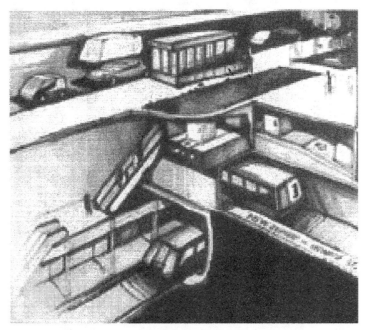

U. of Illinois Chicago tunnel development study concept

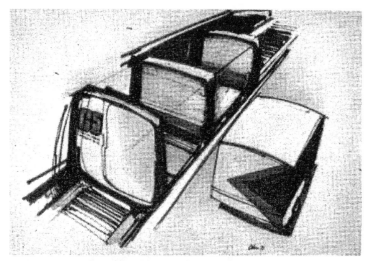

Proposed freight distribution system

U. of Illinois: Automated proposal for existing Chicago freight tunnels

Spare tire is the bumper

The Dragonfly

Hatchback sedan configuration

Touring landau configuration

Open roadster configuration

Touring landau configuration

Open roadster configuration

Utility pickup configuration

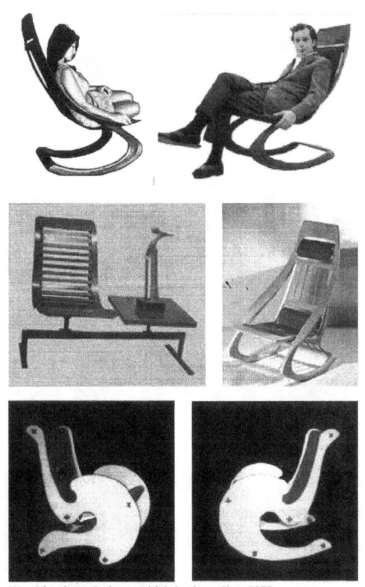

Wood furniture designs and fabrications: Circa 1975

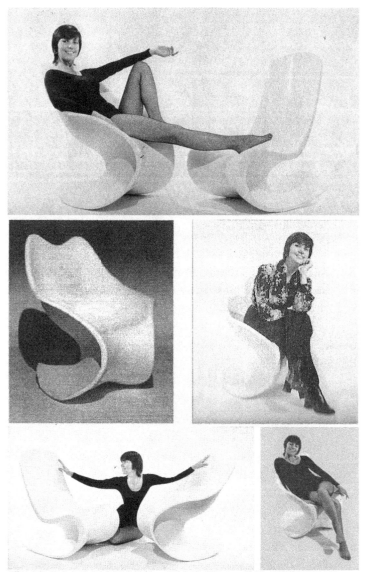

Fiberglass furniture designs and fabrications: Circa 1975

Chapter 5: Pursuing Selfish Passions

FLASHING BACK NOW, the disheartening design compromises involving my first professional architectural project, the Eastern Illinois dormitory complex, taught me a transformational lesson about myself. Whereas my desire to serve interests of others is real, and my desire for earned economic rewards is unapologetic, neither fulfill a larger need.

A large part of my life must be privately reserved for passionate uncompromised pursuit of ever-elusive dreams and ideals—incomplete ideas—tantalizing visions—inspiring potentials—all of these and more.

As I previously wrote in *Thinking Whole,* dreaming may be a forgotten art we must relearn from our inner child-selves in order to rediscover what we truly value most in ourselves and our lives. Assuming that visions direct realities, and I believe they do, then dreaming is something that we can't afford to abandon or postpone.

Maybe visualizing perfect lives—the people we wished to eventually become—came easier in childhood. That was before setbacks and other disappointments challenged our confidence, before we forgot how special we are and before magic was disproven.

Maybe our self-expectations began to wither when parents and teachers warned us about how difficult it is to convert hopes into realities and stressed the importance of looking at life strictly from a practical point of view.

Life is a voyage to meet a new person we are always in the process of becoming. Each leg of our journey offers challenges and lessons that change us. Here, our sense of "being" is a constantly evolving state of awareness and development; an open-ended pursuit of understanding; a perpetual process of "becoming."

George Bernard Shaw observed: "Life isn't about finding yourself, it's about creating yourself."

I believe it is both.

Self-Discovery and Celebration through Sculpture

According to legend, when Michelangelo was asked about difficulties he must have encountered in sculpting his great masterpiece, David, he purportedly replied, "It is easy. You just chip away the stone that doesn't look like David."

Although there is no substantive evidence that Michelangelo actually said that, an essay comment which appeared in a book review of *The Poems of John Greenleaf Whittier* published by The Methodist Quarterly Review expresses this same idea in the context of a broader metaphor:

> *It is the sculptor's power, so often alluded to,*
> *of finding the perfect form and features of a*
> *goddess, in the shapeless block of marble; and*
> *his ability to chip off all extraneous matter,*
> *and let the divine excellence stand forth for*
> *itself. Thus, in every incident of business, in*
> *every accident of life, the poet sees something*
> *divine, and carefully scales off all that*
> *encumbers that divinity, and permits it to be*

revealed in all its transcendent loveliness.[15]

Just as divine goddesses beckon to sculptors and poets from every rock, vast mountains of wonderful possibilities await liberation by those with necessary vision and chipping motivation in every field of endeavor.

Liberating truly meaningful potentials from extraneous slag appeals to my overall design philosophy. It combines big picture references, a quest for logical coherence and a reductive emphasis upon refinement.

As an inveterate carver, I can attest to the fact that liberating even imperfect possibilities from large stones or hunks of lumber typically demands major commitments of time and labor. Sometimes others with whom I share common visions help with the heavy lifting.

I first discovered, or perhaps intentionally initiated, a life-long passion for sculpture as a graduate student in a class taught by Dennis Kowal, a young professor my same age. I had enrolled with a great desire to expand my design interest and background through ambitious works. Recognizing this, after only a couple of weeks later, Dennis suggested that I discontinue attending his class, generously inviting me to share half of his personal studio instead.

Externalizing Internal Visions

Upon sharing Dennis's studio, the two of us almost immediately became professional colleagues and friends, each separately exhibiting our works in gallery shows. I was mostly doing some rather large carved wood pieces which were being

---

[15] *The Poems of John Greenleaf Whittier*, The Methodist Quarterly Review, Whittier's Poems, January 1858, pp. 72-78, Carlton & Porter, DD Whedon, Ed.

shown and represented by the Distelheim Gallery in Chicago…one was purchased by the Illinois State Museum in Springfield. Dennis was working primarily in stone and bronze, with national clientele.

I was gratified that my work was also favorably recognized by other art faculty at the University of Illinois whom I greatly admired and respected. This was particularly significant to me in light of the fact that I was a graduate student, and later, a professor in the Industrial Design program rather than a member of the excellent sculpture department.

One reason that I chose to be represented by a gallery was because many of my works were quite large and I didn't have a place to store them. Several of the wood pieces were about 6 feet tall, a size reflecting a human scale that I could really relate to.

I had decided to display my work the Distelheim Gallery on Oak Street in downtown Chicago because they never attempted to influence my creative priorities. Other excellent commercial galleries that expressed interest in representing my work indicated preferences that I concentrate on certain types of pieces that they thought would appeal most to their particular clientele. I didn't accept their offers because I didn't want to be constrained by marketing considerations that influenced what I would do, how much time I could spend on a piece from a business profitability standpoint or the particular image a gallery wished to project through its stable artists.

My featured one-person Disterheim shows were invariably well attended by the public and Chicago art critics and received a lot of media attention with good reviews in leading newspapers and art publications. The experience of witnessing other people responding positively to products of very personal thoughts and solitary activities was always somewhat unexpected and appreciated, even though such public recognition had little or nothing to do with my private work motivations.

I was also involved in one-man and group shows at other galleries, and a large wood piece was purchased for the prominent collection at the University of Illinois' Krannert Center.

In 1972 I was invited to participate in a broadly advertised and well-attended three-person show at the Illinois State Museum in Springfield. The museum purchased one of my large wood pieces for their permanent collection, where it was prominently displayed at the entrance to welcome visitors.

Material Delights of Making

Although I have produced some cast bronze works using lost wax and investment processes, neither the materials or the methods appeal to me as naturally as carving. Maybe this is partly because modeling enables opportunities to endlessly modify what you are working on, lacking the same level of real concentration, commitment and risk that I enjoy in carving, where each removal action is deliberate and irreversible.

Maybe it is because clay and wax lack the substance and character of wood and stone. And maybe it is because a young child with his first jackknife who could spend hours on end whittling an animal from a fallen branch continues to be part of me.

Most of my earlier pieces were too large to be carved out of solid logs. Even if sufficiently huge trees were available, it would waste a lot of material. Also, natural cracks that formed as the wood cured would weaken or destroy the finished pieces. Instead, I laminated together layers sawed from 2" x 12" kiln dried lumber to provide a mass form only where material would be needed.

The first step was always to do a very rapid sketch that would endeavor to capture and describe the form and spirit that I wished to express. Then, with great care to honor that form and spirit, I produced hard line elevations and top views to visualize where mass would be needed as sections through the piece. Full size patterns were then drawn, enabling shapes to be cut from the redwood or pine planks, fitted and glued together and assembled into laminated layers.

The carving and shaping was primarily accomplished by hand using chisels and files rather than power tools. This is because much of my satisfaction comes from direct and intimate contact with the materials; the nature and direction of the grain; the responses to the tool; the hardness, weight and texture; and the smell. Special satisfactions come from the process of carving, where each cutting or chipping action is an irreversible commitment, and carelessness in calibrating force or angle of a chisel blow can prematurely end a project.

When finished, the thin, light flowing forms exhibited no evidence that they were carved rather than molded. There was an inherently paradoxical anachronistic contradiction in this that probably only a carver can fully appreciate. It seemed doubly deceptive to convey the illusion that such graceful forms might somehow grow out of former geometric slab masses which, in turn, had once been parts of graceful living trees.

I also worked in stone, and still do. On one hand, the cold material lacks the friendly working texture that I find so pleasurable with wood. On the other, the unique varieties of colored graining patterns make each type and piece special... true beauty most fully revealed in a polished state.

My most recent stone sculpture work applies essentially the same lamination approach I developed in assembling the masses from which I sculpted large wood pieces. Here, the individual composite shapes are emphasized by distinctive marble veining patterns and shadow lines that differentiate and define each layer.

Two prototype horses produced by this procedure were assembled from marble slab pieces that were hand cut in

Guatemala using patterns I provided. The process also lends itself to creating works of a large size for public and private settings.

I have also produced some scale models and full-size prototypes for marble pieces that can be fabricated for commercial limited series editions. These are fun to design and nice to see realized in beautiful natural materials. Yet they don't afford the same level of intimate hands-on satisfaction experienced using my familiar mallets, chisels and files to shape them.

I sometimes regret that other interests and responsibilities currently leave little time to pursue major sculpture projects. The large wood pieces that I produced during my active gallery years typically demanded three months of totally dedicated effort that filled annual summer vacation periods. I have, however, indulged subsequent urges to sculpt some smaller figurative pieces out of walnut and bass wood, several later-to-be-discussed bears included.

Child's Play

I don't overly concern myself about whether or not some of the things I choose to make for personal reasons qualify as "art." Is a particular representational depiction of a literal object or event merely illustration? Should an abstraction that happens to be visually appealing be dismissed as only decorative? Is a created work that offers insightful social commentary more or less relevant than another that eloquently expresses form, space, balance, movement and structure? Can a work created for fun also be taken seriously?

Such momentous questions are for others to ponder and debate. Meanwhile, I'll allow myself to be content doing whatever interests me.

Some years ago, I became fascinated by the beauty and elegance that characterizes fine Inuit carvings produced in high northern latitudes. Perhaps the creators of these pieces as well as some modern critics would not distinguish them as art. Yet they often exhibit a wonderful combination of sophisticated attributes, including cultural narrative, innovative expression, an intimate understanding of materials, a refined sense of form and proportion and superb craftsmanship. The great sculptor Henry Moore recognized and respected these qualities.

Various travels to Greenland, Canada and the Japanese island of Hokkaido provoked a special interest in the life views and traditions that influenced these accomplishments I deeply admire. Cultures in all of these regions have recognized strong spiritually and physically interdependent relationships between humans and other creatures that share their natural environments.

Bears have been traditionally revered as special spirit beings, which partly explains why they are frequently featured as subjects of expression. For reasons that are not fully clear to me, I began to carve lots of bears myself. Most were polar bears. I marvel at their ability to survive, even thrive in unimaginably harsh and desolate places.

I experienced a great honor and thrill when some Japanese friends arranged for me to visit an accomplished bear carver in a remote Hokkaido village. We spent a wonderful day working together on a piece that we took turns carving, passing the unfinished bear back and forth between us over many silent hours.

As he made a few cuts on one side, I would carefully duplicate those same features on the other as the form slowly emerged from our collaborative shaping. Although we shared no spoken language, our communication of everything important required no words.

The unfinished bear was then sent to me at my home where I completed it. I hope that my masterful teacher would approve of the finished results.

My Crafty Side

I have always enjoyed drawing and became interested in graphic and jewelry design activities in college, but I have never practiced these craft pursuits on a commercial basis.

Good drawing and graphic skills which were emphasized in both my undergraduate architecture and graduate industrial design training came in handy in basic 2-D and 3-D drawing and composition I taught at Parkland College. Some of my students later became successful professional artists and design professors.

My personal art, illustration and graphic work has been pursued strictly for pleasure and my own corporate business purposes. Having founded numerous organizations, there have been many opportunities to create graphic logos for products and identity programs.

Jewelry design and fabrication has afforded me means to indulge my enjoyment of conceptualization and craft on a personal level. Some pieces have combined carved wood and metal elements, some have been hammered or bent from silver plate, and some were carved out of wax and cast. An advantage of this hobby is the ability to create a piece in only a few hours or days, making the necessary time commitment much more manageable within my schedule than sculpture or most other projects.

Although the jewelry forms I create often resemble certain aspects of my sculpture, I never conspicuously plan for this to occur.

While technology is continuously yielding wonderful new tools and toys that enable us to visualize, present and fabricate things with high fidelity and speed, I regret that this trend may be rapidly eroding general appreciation of the special beauty and satisfactions associated with fine human craftsmanship. I am certainly impressed and even awed by new dimensions of design exploration and communication afforded by computers, and particularly by computers' animation and virtual reality applications. But I do not believe that knowledge and skill in these areas offer a substitute for prerequisite sketching abilities that are essential to idea formulation, development and examination.

I greatly admire artisans who have dedicated their lives to perfecting a craft. Having been a carver throughout my childhood and adult life I have a particular affinity to wood block printing and metal etching, although I have never actually attempted to perform engraving.

Examples that relate to a special interest of mine are fine engravers throughout history who have applied their skills to guns, knives and swords as "metal canvases." This is enormously time consuming and exacting work that involves the use of chisels to achieve marvelously elaborate and detailed results, sometimes incorporating inlays of gold and other precious materials.

I commissioned one of the best contemporary gun engravers to execute a design I created for one of my favorite types of pistols, the venerable 1911 Colt. This is the same firearm that some of your great grandfathers carried into battle during World War I. Now, nearly 100 years after the genius of John Browning first created this masterpiece of design innovation, it still remains virtually unchanged as the most popular choice for competition match shooters by a large margin.

The gun I sent to the engraver was a 1970s vintage example. The process, based upon my drawings, took him about a year to complete. One of the most accomplished U.S. ivory carvers produced the grips I designed to complement some of the gold inlay images on the gun's metal slide.

Sculpture and Illinois State Museum exhibit: Circa 1970

Carved wood sculpture

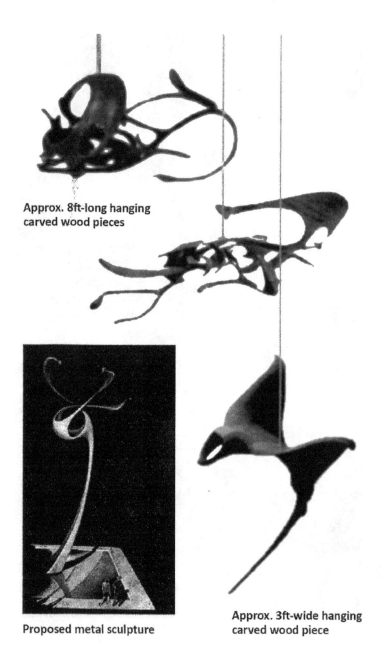

Approx. 8ft-long hanging carved wood pieces

Proposed metal sculpture

Approx. 3ft-wide hanging carved wood piece

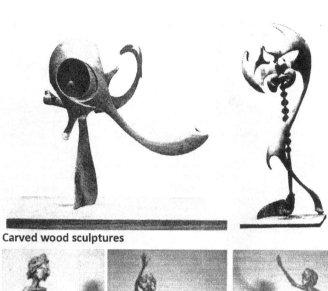

Carved wood sculptures

Bronze sculptures

Stone sculptures

Fabricated stone sculptures apply the same process used to laminate bulk forms for large carved wood pieces

Fabricated marble sculpture (full-size horse)

Wood pattern for stone bird

Wood models for fabricated stone sculptures

Wood model for large outdoor stone sculpture

Larry Bell

Wood model for large outdoor stone sculpture

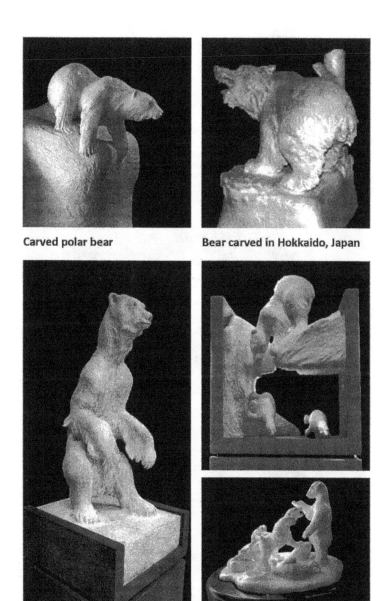

Carved polar bear

Bear carved in Hokkaido, Japan

My polar bear carving phase

Carving of myself

Nancy with carving

Baby Aaron with rocking horse

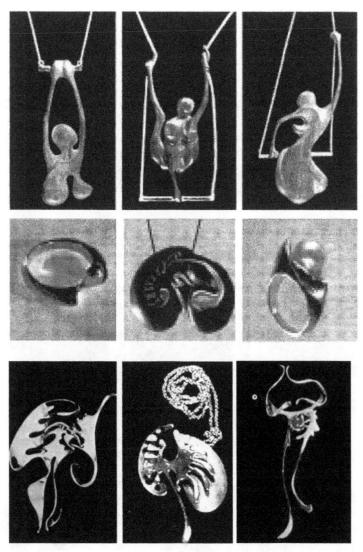

Jewelry designs

Miscellaneous images

Wood cut prints

Chapter 6: Crime Prevention Through Environmental Design

DISTURBING IMAGES OF urban decay—a world described by Jane Jacobs and John Steinbeck—abruptly revisited my psyche in connection with an unexpected April 1974 telephone call. A senior partner with Barton-Aschman Associates, Inc. (BAA), a large transportation engineering and urban planning firm, asked if I would be interested in heading their company's project regarding a major national research and demonstration program addressing environmental and design influences upon crime in various settings.

As with many other opportunities in my life, the invitation had resulted from a serendipitous chain of seemingly unrelated planning and design experiences. BAA's office in Evanston, a Chicago suburb, had previously hired me as a consultant to design and produce some street furniture for one of their urban projects. Some of those elements were molded from fiberglass by Steve Norcross. BAA was pleased with the end-products which were produced on time and under budget.

The BAA principal told me I immediately came to mind for new opportunity which called for a "maverick" planner and designer with the willingness and ability to jump into uncharted waters…they decided that I might be their guy. The work would require me to take a faculty leave of absence from the University of Illinois, temporarily relocate and be based out of their Washington, DC office, and engage in extensive travel throughout the United States.

I immediately replied: "of course."

CPTED Program Background

BAA had become the planning and design consultant to the Westinghouse National Issues Center in Arlington, Virginia. The center had won a multi-million dollar "Crime Prevention Through Environmental Design" (CPTED) program contract issued by the U.S. Department of Justice—Law Enforcement Assistance Administration (LEAA). The CPTED initiative grew out of studies conducted by an architect, Oscar Newman, who had examined relationships between crime problems and the design of New York public housing projects. His 1972 book, *Creating Defensible Space,* had captured the attention of top administrators in the criminal justice area.

Newman's book advanced a variety of influential observations and theoretical conclusions. The author noted and attributed higher crime rates in high-rise buildings than lower projects to residents' feeling a lack of social control or personal community responsibility—a sense of territoriality—for areas occupied by larger numbers of people.

As Newman described a responsive defensible space priority must then be:

> *To release the latent sense of territoriality and community among inhabitants so as to allow the traits to be translated into inhabitants' assumption of responsibility for preserving a safe and well-maintained living environment.*

Newman argued that residential areas in public housing projects should be subdivided into smaller entities of similar families to enhance both an actual and perceived sense of inhabitant control. He observed that smaller groups, families in particular, more frequently use an area geared for them, and as the number of activities increase, a feeling of ownership and need to protect the property follows. On the other hand, when larger groups use a community space, no one has control over the area leading to conflicts over acceptable uses and activities.

Oscar Newman also emphasized a priority to design buildings and spaces in ways that enable natural surveillance, which enables residents to be watchful for intruders. This need applies to planning on a neighborhood scale and in the design of building complexes and structures. Both applications can either discourage or encourage resident safety and sense of safety through such considerations as lighting, adjacent area uses and activity levels and visual openness for problem avoidance and detection.

Other researchers and writers echoed similar conclusions. Jane Jacobs, for example, had previously noted that city streets often lack three primary qualities needed to make them safer: clear demarcation between public and private space, multiple uses, and a vitality of activity which optimizes the number of "eyes on the street."

Westinghouse National Issues Center Program

Within two years of Newman's book, the DOJ-LEAA awarded a competitive contract to the Westinghouse National Issues Center to study crime problems and implement interventions in three different types of environments: a commercial urban strip, public high schools and a private middle-income residential neighborhood. Of these, my roles concentrated intensively upon the first two categories which presented the most urgent crime and intervention challenges.

Tom Reppetto of the John Jay College of Criminal Justice

and James Tien with Urban Systems Research and Engineering, Inc. headed the crime data analyses responsibilities, and Wiles BAA was retained to plan and implement the demonstration projects.

It is highly relevant to note here that all financial and public service resources necessary to actually test and demonstrate our proposals were required to be voluntarily contributed through public/private funds and in-kind contributions originating from within the targeted communities. This meant that we needed to sell the communities on good ideas based entirely on merit.

The central purpose of the Westinghouse CPTED program was to identify and demonstrate countermeasures to contributors of the physical and social environments that provide opportunities for or precipitate criminal acts. In doing so, the intervention strategies were also to address and mitigate fear of crime, build community cohesion and pride and advance quality of life benefits.

Unlike more typical urban crime prevention programs, our CPTED approach turned the strategic emphasis away from target hardening—the use of artificial barriers, walls, fences, gates, locks and other tools to impede accessibility. I have often irreverently referred to this as a "moats and alligators approach" which only displaces the problems...sending the culprits to more vulnerable targets next door.

Portland, Oregon Commercial Demonstration

Our Westinghouse team selected a declining three and one-half mile-long, four block-wide mixed-use area known as the Union Avenue Corridor for a Commercial CPTED demonstration. Once a thriving community along a major transportation artery, by the late 1950's, a new north-south state highway system routed traffic and business markets away.

Racial disturbances in the late 1960's and early 1970's

contributed to further urban decay, increased predatory crime and burglaries and led to pervasive community fear.

Although conditions had since stabilized, lost business and urban neglect had left the community with a fair-to-poor condition mixed use of light industry, new and used car dealerships, grocery and variety stores, and residential housing. Remaining merchants had little confidence about the area's future, and an anchor grocery establishment was planning to leave "back-door" due to costly theft losses.

It was immediately clear to us that community reversal initiatives would require crime prevention, economic development and social revitalization strategies that needed to be planned and implemented as common, interdependent priorities. This required close and active collaboration with local government leaders and planners to obtain and strategically leverage vital public funding and support services.

Key CPTED priorities were to energize local leadership in proactive engagement with municipal officials and entities, to retain and expand commercial and public services neighborhood residents depended upon, to reduce crime risk and fear for most vulnerable citizens and to strengthen neighborhood identity and investment confidence essential for progress.

And not to be forgotten…to accomplish these goals with no money allocated on our behalf for project implementation.

We worked with Portland's public works department to target street lighting improvements following prioritization of this need by the local merchants' association. This strategy had an obvious crime prevention purpose, but perhaps equally important, helped to establish closer cooperation among local businesses, and simultaneously, between the business association and municipal leaders.

We helped to establish and coordinate security and community enhancement with other municipal agencies as well. For example, building address numbers were placed in alleyways behind commercial establishments to assist police in identifying and responding undetected to locations where break-ins and robberies are reported.

Merchant establishments were encouraged to remove store window advertisements and displays that blocked police surveillance visibility of interiors at night. Security advisement services were provided, including target hardening through upgraded locks, cameras and alarm systems.

A "Safe Streets for People" initiative provided mini-parks with lighting, emergency telephones and open bus stop enclosures along the corridor. Bus transit schedules were upgraded to reduce waiting times—therefore reducing periods of crime threat exposure—particularly at night.

Purse snatch incidents represented a prevalent and especially traumatic problem for elderly women who were susceptible to serious, potentially fatal hip-fracture injuries resulting from being pushed and falling in the process. In response, we worked with local banks to establish a "Cash off the Streets" (COTS) initiative to encourage and teach residents to open checking accounts rather than depend upon currency transactions. Subscribers were given COTS buttons for attachment to clothing and handbags to discourage potential predators.

A Union Avenue Corridor promotion program was launched to attract additional private investment and services. Although not implemented, I proposed and designed the development of a small commercial retail center as part of this campaign.

Altogether, the broad level of community cooperation realized through this unfunded demonstration struck me as remarkably encouraging.

Florida Broward County School Demonstrations

Four high schools in Broward County, Florida were selected for CPTED school demonstrations. They were intentionally picked among dozens of others to represent a range of economic and demographic populations, student academic performance ratings and social behavior challenges:

- Boyd Anderson High School was bordered by racially-mixed neighborhoods and represented low to lower-middle income families. Student academic performance was well below average. Principal problems included individual and group assaults, burglaries and extortions by students against other students on school grounds, property theft, school vandalism and graffiti.

- Deerfield Beach High School served a predominately middle-income population in a racially mixed residential area. Student academic performance standings were below average. Most behavior problems, including assaults, robberies and extortions occurred outside the immediate school premises, in parking lots.

- South Plantation High School primarily served a middle-upper class neighborhood. About 79 percent of the residents were white, and overall student academic achievement ranked above average. Behavior problems were generally individual and relatively limited.

- McArthur High School served a middle-upper class neighborhood, about 85 percent of the students were white. Student academic performance was above average, and behavior problems were moderate.

Although each of the four schools were different from one another in many respects, all shared similar problems which influenced common CPTED goals.

A common priority, the most difficult of all, was to constructively influence the institutional culture to improve levels of mutual trust and respect between administrators, teachers and students. The goal was for problem-aversive administrators and teachers to act less like prison wardens, and for students to behave less like prisoners conspiring to plan jail breaks.

A common response to bad student behavior was expulsion, transferring the individual and the problem to the streets. A more ideal strategy would be to make expulsion a penalty rather than a reward for disruptive behavior by making the school social and activity environment more interesting and exciting.

It is naïve to imagine that such attitudinal transformations can be expected to magically come about through physical changes by others to the school environment, however wonderful those changes might prove to be. A much-preferred strategy, as I saw it, was to somehow engage students and teachers as the primary social and physical change agents; to engender a sense of accomplishment and pride; and to make schools not only safer, but more experientially and intellectually stimulating as well.

Great credit is owed to administrators, teachers and students who put such principles into practice. An art teacher at one of the schools engaged her students in painting large topical super-graphic wall images to identify various classroom and activity areas. The highly professional results dramatically transformed drab settings into environments of vitality and pride.

Students under leadership and supervision of an instructor in a shop class designed and constructed furniture for an enclosed outdoor courtyard area. Small tables replaced previous larger ones to encourage safer interpersonal socialization rather than attract disruptive group-think behaviors.

The wonderful results of these projects brought a sense of positive identity and well-being not only to the students and faculty, but to the administrators and community as well. And yes, a number of more conventional security initiatives were implemented as well, including improvements to indoor and outdoor lighting to enhance natural surveillance.

Altogether, including the Portland Union Avenue Corridor initiatives, I believed that we had accomplished productive and satisfying achievements.

Personal Civics Lessons

Life in the Washington, DC area introduced me to another life-long world of interest...a discovery of purposeful civic responsibility and political policy engagement.

Cindy and I were living in Reston, Virginia, a planned community located about 25 miles west of Washington not far from the Dulles International Airport. Founded in 1964 by developer Robert Simon, Reston was one of first "new towns" influenced by the Garden City movement which emphasized self-contained communities that intermingled green space, residential neighborhoods and commercial development.

Our leased home was located in Lake Anne, the first of Reston's villages. The village center actually does have a small artificial "lake" immediately adjacent to commercial shopping and businesses. It was a lovely environment.

At that time, during the mid-70's, the typical 40-minute commute to and from my Washington office at 18th and K Street to Reston was pleasantly scenic. The route which then passed through green woodland and horse ranches has since become dominated by residential suburbs, and I'm told that the travel time has at least doubled. Reportedly, high-density commercial and residential developments along the Dulles Toll Road have put a strain on the local governments' ability to ensure that key infrastructure, including service roads, schools and parks remain in synch with the pace of new construction.

Having said this, the natural beauty of Northern Virginia combined with the Washington-area's civic vitality and rich historical legacy continues to draw me back. I experience an affectionate and nostalgic sense of being home upon each return.

The human scale expressed throughout planning and architectures of our nation's capital befits a country that celebrates individualism. Streets everywhere are full of life: people rushing to go somewhere to participate in something they believe is urgent; families and teachers introducing children to world-changing places and processes; joggers dodging dog-walkers and stroller-pushing parents along placid Potomac paths; and altogether, professional and personal experiences of purpose and pleasure that share springtime cherry blossom fragrances.

While in Washington pursuing CPTED, I also attended the U.S. Federal Executive Institute (FEI) in Charlottesville near the University of Virginia. Although established to exclusively serve top-tier government "supergrade" civil service leaders, I was one of the quite rare non-government participants they occasionally invited to introduce outside-the-Beltway experiences and perspectives.

The FEI offers advanced interagency leadership development programs and curricula which allow "civilian generals" to meet, exchange ideas with and learn from ranking counterparts from across government. Many of the activities are planned and structured to encourage participants to look seriously at themselves and their careers to consider what changes are warranted and what core personal resources they need to bring those changes about.

I gained many valuable insights from FEI self-examination exercises that I have since applied with my own students. One example engages individuals in very small groups to identify personal challenges and obstacles they have overcome in their personal and educational lives as others query them about strengths and strategies they employed to succeed. Each member of the group takes turns being the trusting revealer and an attentive inquirer.

Applying CPTED Economic Development Principles

I continued to pursue CPTED consulting upon returning from a University of Illinois faculty leave of absence.

Although progress towards reducing crime incidents and fear at all Westinghouse demonstration sites seemed encouraging, I sadly had no illusions that our short experiment had really even begun to solve the core problems.

Nevertheless, fundamentally logical, CPTED lessons continue to guide criminologists and security planners in cities throughout the United States and world. Many police departments retain professionals who are specially trained and assigned to these programs. The U.S. Department of Housing and Urban Development applies many of these basic principles as safety and quality-of-life enhancement strategies.

It soon dawned upon me while pursuing CPTED school demonstration consulting in Florida that some preferred thermal aspects of my residency choice were woefully out of sync. Somehow, I inexplicably found myself returning from sipping margaritas outdoors in a balmy winter location to again inhabit a frozen and windy Champaign-Urbana prairie.

The University of Illinois had been a wonderful experience. Despite that advance resignation letter the dean had asked me to sign years earlier, I had been promoted to full professor during my year-long CPTED absence.

Still, it struck me as high time to move on to fresh

experiences and adventures in far warmer places. Dispiriting dreams about snow one day drifting across my tombstone in Champaign-Urbana contributed to that motivation to move on. And although I had never before been there, Texas seemed to be a good place to try on for size.

So, I did.

A road trip south with some teaching interviews along the way landed me a professorship position in the College of Architecture at the University of Houston.

Soon after arriving at the U of H, I established an organization called the Environmental Center: Houston within the College of Architecture, to conduct funded urban research and planning. Much of this work focused CPTED-related principles and strategies upon economic and environmental revitalization of neighborhoods surrounding and adjacent to the university central campus.

One initiative called *Incentives for Inner City Revitalization* was sponsored by the Mayor's Office, the U of H and the neighboring business and community organizations to attract needed businesses and services to distressed neighborhoods. A key achievement of the initiative was the establishment of a very active and effective East End Progress Association to represent and advance local improvement priorities.

Another Environmental Center research study sponsored by the U of H Board of Regents addressed a broad range of campus and area-wide neighborhood infrastructure and institutional policy issues. Major emphases were to improve the surrounding image and quality of life through improvements that would offer better housing options for faculty and students, attract restaurants and other commercial services and reduce crime and fear problems. Many of these proposals guided changes to campus highway connections, numerous land purchases and upgrades and new student housing developments.

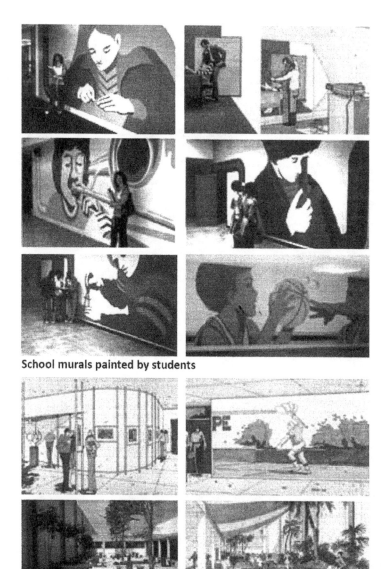

School murals painted by students

Crime prevention through environmental design in Florida schools

Proposed commercial center design

Conversions of street areas to pocket parks

Area lighting improvements
Crime prevention through environmental design, Portland, Oregon

Chapter 7: Architectures Beyond Earth

SOON AFTER I established the Environmental Center in 1978, I began to explore other interesting planning and design opportunities in the Houston area. The NASA Johnson Space Center was a natural place to start my search because space presented very interesting problem-solving challenges. Some JSC organizations were just then beginning to think about requirements for a space station, although the program had not yet been formally approved by Congress or NASA Headquarters.

I managed to secure a small NASA contract which was shared with the U of H Hilton College of Hotel and Restaurant Management to study menu requirements and living accommodations for space station crews. This work led to many other design studies, including an orbital structure concept we developed with Apollo astronaut Buzz Aldrin, who has become a very close and lasting friend.

Guillermo Trotti soon joined me in these activities. "Gui" had first worked on space design projects as an undergraduate student at U of H and continued to work in this area as a graduate student in architecture at Rice University. Together, we introduced the topic to students here in connection with a new Experimental Architecture studio.

Larry Bell

A Higher Design Frontier

Science fiction began to merge with reality-based aspirations that space was a logical frontier for exploration and development by current or near-term generations of pioneers. The idea of living in weightless environments where "up" and "down" were determined only by design challenged our imaginations, including some UH architecture students who wished to work with us.

Whereas space design had generally been associated somewhat exclusively with engineering and government astronauts, these assumptions gradually seemed to be changing. The idea that architects and other designers might soon be important contributors seemed unprecedented. So, we simply paid no attention, and did it anyway.

Space Architecture—as we apply that term—entails research, planning and design of space missions, orbital vehicles and habitats and living and work accommodations on lunar and planetary surfaces. By necessity, it draws upon knowledge and skills that are associated with many professional disciplines, including architecture, engineering, industrial design and life and physical science fields.

The activity requires a good working understanding of a variety of conditions and requirements that are uniquely present in space environments. Examples include: influences of weightlessness or reduced gravity levels upon health and operations; radiation hazards and interventions; and severe influences of launch and landing constraints upon allowable mass, volume and many other aspects of design.

The early Space Architecture projects we undertook rapidly began to gain real prominence and credibility within the Houston aerospace community. This led to larger to larger U of H Experimental Architecture studio research and design contracts, which in turn opened up opportunities for our graduating students to secure employment at various NASA centers and major aerospace corporations. Expanding activity attracted more and more highly motivated students to get involved.

Some of the early concepts we proposed received widespread national and international interest which continues even today. A good example is an inflatable space facility we named Spacehab. This design approach enabled large, flexible pressurized structures to be collapsed to fit in the cargo bay of the then-existing Space Shuttle Orbiter for launch, and then be deployed in space.

A Transhab module based upon this same inflatable approach was proposed and pressure tested by the NASA Johnson Space Center. Much of its design was undertaken by one of our U of H architecture graduates who worked on the concept with us prior to his employment at NASA.

Bigelow Aerospace has since developed some working inflatable prototypes, including a full-size habitable module attached to the International Space Station. I provided research and design support to Robert Bigelow when he first initiated his company's enterprise.

Our Spacehab concept, which was proposed at a time of growing international interest in space exploration, generated a tremendous amount of international public attention. Popular interest in our space architecture work was particularly keen in Japan where it was featured in numerous major newspaper and magazine publications.

As a result, I was invited to visit Japan on dozens of occasions to appear on television interviews and to present lectures to aerospace and academic audiences. I also talked about space architecture at public and technical forums as well as at entertainment events. Sometimes this was in the company of U.S. astronauts and representatives from Japanese space organizations.

One trip involved an extensive lecture tour with U.S. shuttle astronaut, Dr. Joe Allen. Joe later became the

Larry Bell

president, CEO, and finally Board Chair of Space Industries International, a commercial space company I co-founded which is discussed later.

I was invited to present a keynote address at the 1985 International Housing Symposium that was held in Tsukuba, where participants came from all over the world. I also had the privileges of attending the first launch of a new rocket at Tanegashima Space Center on the island of Kagoshima; to join with the Mayor and other officials from Narita City in dedicating a new Space Science Museum; and to give a speech to municipal leaders in the capital city of Sapporo in 1987. The following year I was honored to receive the Space Pioneer Award from Kyushu Sangyo University in Fukuoka.

In addition, I participated as a member of a U.S. delegation that spoke at a large "Space Decade '90s" conference in Tokyo, and I was a keynote speaker at a Space Business Conference that took place there that same year.

A big ego-inflating highlight, of course, was a feature about me which appeared in a Japan issue of *Playboy Magazine*...although perhaps mercifully to readers, not on the centerfold.

Sasakawa International Center for Space Architecture

If you think that this celebrity stuff was going to my head, I won't entirely deny that. But my other purpose for mentioning the Japanese attention is to illustrate some developments that led up to a $3 million gift that established the Sasakawa International Center for Space Architecture (SICSA) in 1987. Many people ask how that came about, and often seem to assume that this is when our unique space architecture program began. In reality, the endowment actually happened after, and most certainly because, our program had already become well known and respected throughout Japan.

I was introduced to Ryoichi Sasakawa by astronaut Col. Gerald Carr, the Commander of Skylab IV, during a shared lecture tour throughout Japan in 1985. Gerry had been invited to meet Sasakawa following his historic flight, the longest U.S. space mission. Mr. Sasakawa was an extremely famous and wealthy man who contributed many millions each year to philanthropic causes. For example, he was the principal sponsor for the former President Carter's hunger relief activities in Africa and is credited with funding the medical programs that have eradicated leprosy as a major world health problem.

During the course of our discussion, I suggested the possibility of creating an international center devoted to peaceful uses of space and space technology. Mr. Sasakawa expressed interest and requested a written proposal. I responded and, following a visit here by his representative to review our program, one of his organizations, the Japan Shipbuilding Industry Foundation (JSIF), provided a $3 million gift. The U of H matched that endowment through the dedication of the 3rd Floor East Wing of the College of Architecture building for SICSA.

People sometimes ask if there are any strings connected to the generous Japanese gift. Other than wanting assurance that the U of H will support SICSA in making it a world-class center, the answer is no.

Mr. Ryochi Sasakawa, who passed away in 1995 at age 96, was succeeded by his son, Yohei Sasakawa, as chairman of JSIF, which has since been renamed Nippon Foundation. I had originally met Yohei during early SICSA gift arrangement discussions, and I was greatly pleased and honored to renew our acquaintance during his visit here in 2018. Those 30 years since SICSA was first established have been enormously eventful and gratifying, effecting the lives and careers of hundreds of our space architecture graduates from countries throughout the world.

Larry Bell

Addressing Extreme Environments Everywhere

A guiding purpose of SICSA is to advance peaceful and beneficial international development and uses of space and space technology, a goal that applies to terrestrial as well as space environments. Accordingly, much of SICSA's emphasis also addresses comprehensive aspects of research, planning and design for a wide variety of extreme settings on Earth. Examples include surface and underwater ocean science and commercial facilities, polar research stations and shelters for victims of natural and manmade disasters.

Some of these extreme Earth environments present useful analogs for planning and designing missions and facilities in space. Some common issues include isolated locations which limit access, present special construction challenges and impose harsh and dangerous living and working conditions. All require careful planning to optimize human safety and self-sufficiency.

Useful concepts, lessons and technologies are often also transferable between different types of terrestrial environments. An NSF-sponsored SICSA study design proposed using jack-up ocean oil rig structures in combination with a special wind-tested aerodynamic architecture to prevent polar research stations from becoming buried by drifting snow. Some of our graduate students traveled to Greenland to install a scale model mockup demonstration of the concept.

SICSA co-founded the International Design for Extreme Environments Association (IDEEA) in the late 1980s, along with cooperating organizations in Russia and Canada. IDEEA-USA, headed by SICSA, then planned and hosted a large week-long conference (IDEEA One) that was held on the U.S. campus in 1990. More than 400 professionals from diverse fields participated, representing 12 countries.

IDEEA One was supported by NASA, the National Science Foundation, the U.S. Bureau of Mines and the U.S.

Army Construction Engineering Research Laboratory. Other government and corporate organizations contributed in-kind services and offered official endorsements. Featured speakers included Dr. Sylvia Earle, who has set ocean deep diving records, and Will Steger, who headed a small team that walked across the continent of Antarctica.

The IDEEA activity began when I was invited to bring some of my SICSA students to Moscow to tour Russian space technology and science organizations shortly after the collapse of the USSR. During our visit our hosts arranged a special workshop that involved officials and experts from many types of extreme environments.

In addition to cosmonauts and space engineers, they included: people who had been responsible for cleanup and relocation activities following the Chernobyl nuclear disaster, polar researchers and construction experts and medical specialists, to name only a few. Well-known architect Paolo Soleri who established the Cosanti Foundation and Arcosanti urban laboratory in Arizona also attended from the United States. IDEEA One was succeeded by IDEEA Two in 1992, which took place in Montreal.

Some SICSA projects connect surface and space architectures and operations more directly. For example, we have conducted funded spaceport development feasibility and facility planning studies for the Texas Aerospace Commission as well as for the Houston Airport System, both of which foresee a robust commercial space launch future.

Outreaching Globally

SICSA routinely collaborates on joint research, design and educational initiatives with institutional, business and government entities worldwide. One example entails co-hosting ongoing graduate-level summer aerospace workshops in Russia in cooperation with Bauman University (Moscow) and Rice University (Houston).

SICSA was a founding organization in the establishment of the International Space University (ISU) in April 1989. ISU originally began as a private non-profit educational enterprise that organized post-graduate-level three-month workshops in different countries each year taught by faculty from leading international government, corporate and academic aerospace organizations and departments.

ISU's summer programs offered multi-disciplinary lectures and space-related planning and design problem-solving project challenges. SICSA sponsored and taught the space architecture-related disciplines. I led these activities for sessions which took place in Cambridge, Massachusetts; Strassbourg, France; Toronto, Canada; Toulouse, France; and Kitakyushu, Japan. In 1994, ISU established a permanent campus and faculty in Strassbourg.

ISU was co-founded by three visionary young fellows: Todd Hawley, Robert Richards and Peter Diamandis. I first met Peter when he was a graduate student at MIT. In addition to ISU, he later established the $10 million X Prize (subsequently renamed the Ansari X Prize) which was awarded to Microsoft co-founder Paul Allen's successful October 4, 2004 launch of his experimental spaceplane SpaceShipOne. The spaceplane was designed by my friend Burt Rutan.

In 1993, I created the Global Future Foundation, a non-profit organization that raised awareness and money to support our educational programs at SICSA. It accomplished this over a 15-year period by holding annual multi-day fundraising social events in Manhattan. These multi-day festivities included private cocktail and gallery parties, culminating with a featured "Crystal Ball" black tie dinner gala held in elegant hotel settings. Several were held in the Plaza Hotel's Grand Ballroom and at the New York University Club.

The Crystal Ball and its affiliated events were organized by "Brownie" McLean, a prominent Palm Beach, Florida figure. Brownie's social influence drew hundreds of attendees from across the United States and other countries to each gathering. A key purpose was to promote widespread awareness about SICSA, and all net proceeds went to the Global Future Foundation to support our programs.

My very dear friend Brownie, a strong advocate for space and SICSA, celebrated her 100th birthday in 2017. She is one of the very finest and most generous people I know.

Space Architecture Earns Academic Credit

In 2000, nine young engineering professionals from NASA and aerospace firms asked if it might be possible for SICSA to establish a Master of Science in Space Architecture degree-granting program which they could enroll in on a part-time basis while retaining their current working positions. We all realized that in doing so, the curriculum development would need to be closely coordinated with their aerospace employers. Any such program would also require approval by the University of Houston and Texas Higher Education Coordinating Board.

Those nine people not only helped to plan that program, but in 2002-2003 became its first class of graduates...the world's only academic degree-granting space architecture institution.

In 2014, SICSA's management administration and scholastic programs were transferred from the U of H College of Architecture to the U of H Cullen College of Engineering, a move which reestablished MS Space Architecture courses and credits within the college's broader Mechanical Engineering and Aerospace Engineering curricula. In 2017, a joint MS Aerospace Engineering-Space Architecture degree was also established.

Upon SICSA's transfer to Engineering, I cheerfully passed the baton of its leadership as director to former NASA 5-Space

Shuttle-mission astronaut Dr. Bonnie Dunbar who then headed the Aerospace Engineering program. When Bonnie relocated to Texas A&M University in 2016, Dr. Olga Bannova, a MS Space Architecture graduate who had worked with me for over a dozen years, replaced her as SICSA's director.

Many students in our SICSA programs are employed at NASA and major aerospace companies who typically pay their tuition expenses. They come from many different technical areas, including aerospace engineering, mission operations, math and physics and a variety of science disciplines. We also attract many international students with undergraduate degrees in architecture and other fields.

I continue to teach in these programs and share pride with my colleagues in a long history of progress. Dating back more than three decades, many of our former students have pursued careers activities which were very similar to, and sometimes directly resulting from, research, planning and design work they were first introduced to here.

It is sobering for me to realize from a warp-speed time perspective that one of these people, Kriss Kennedy, retired from his space architecture position with NASA as an advanced mission planner this year. Kriss, along with Larry Toups, another SICSA space architecture alum and now retired NASA employee, have also taught as adjunct professors in our MS Space Architecture/Aerospace Engineering-Space Architecture programs.

I believe it is worth special mention that SICSA and its programs have always been—and continue to—be financially self-supporting through several millions of dollars generated from outside research contracts, grants and gifts. These independent resources provide faculty salaries, graduate research and teaching assistantships and part-time student employment on funded NASA and aerospace company projects. In addition, these monies purchase computers, furniture and office supplies.

Many of our students secure part-time employment with aerospace companies—activities that typically draw directly upon and apply their space architecture expertise. These experiences often serve as gateway opportunities for post-graduate positions with the same organizations.

Practicing What We Teach

Flashing back in time now to the late 1980's, Gui Trotti and I formed Bell and Trotti Inc. (BTI), a space architecture consulting company that provided research, design and even full-scale mockup fabrication services to many of the major government and corporate aerospace organizations. Our clients included the NASA Johnson Space Center, the NASA Marshall Space Flight Center, Boeing, Martin Marietta, Grumman and ILC Space Systems.

BTI grew to employ about 30 full-time professional designers and fabricators. All of our design staff, typically about 12, were former students we had worked with in our University of Houston classes...all top-rate talents. Our fabricators were highly skilled at producing just about anything, from very precise small-scale models to whole space station module replicas with detailed interiors.

One such module was developed for Boeing and trucked to the NASA Marshall Space Flight Center in Alabama. A different one was designed, constructed and shipped to Tokyo for a special public space exhibit.

Much of our work focused upon requirement analyses, design development and concept assessments for Space Station Freedom that NASA was proposing at that time. Space Station Freedom, envisioned primarily as an American national initiative, was later superseded by the International Space Station (ISS), which is still operating today.

BTI initially supported three companies that were competing with one another to become the primary Space

Station Freedom contractor: Boeing, Martin Marietta and Grumman. We reported at top management levels of all three and realized as competition became very fierce that we would have to limit our services to one.

We decided to go with Boeing for two reasons. First, our scope of work for them was the broadest, involving configuration planning for the entire station, layout and design of all modules and detailed design of crew systems and other elements. Second, they were very responsive to most of our proposals. Our decision proved later to be a good one, as Boeing won the competition.

I think that it is reasonable to believe that we had made significant contributions to Boeing's successful space station contract bid. In addition to undertaking most of their crew system design and building many models and mockups which we delivered to Huntsville, Alabama, we also produced a number of illustrations depicting our concepts that were highlighted in their winning proposal to NASA.

Our research and design studies covered virtually all major crew support facilities, including galley food preparation and wardroom equipment, toilet and hygiene areas, emergency medical systems and sleeping quarters.

BTI conducted extensive galley studies and designs for ILC Space Systems and supported them in developing a working prototype. We also addressed requirements and options that applied to the overall space station architecture, including interior layout and utility system trade studies, pressure hatch concepts and viewport design and integration. This work was featured in Aviation Week, the leading U.S. aerospace industry magazine, as well as appearing in many other national and international publications.

Our Bell and Trotti firm also undertook architectural projects. Some examples are a competition proposal for a large science center in Riyadh, Saudi Arabia; a lunar settlement set design for a TV drama titled *Plymouth*; and a relocatable space theme park plan we developed through a contract with Kenneth Feld, the creator of Siegfried and Roy as well as other well-known entertainment productions.

The IDEEA One Conference

IDEEA USA President Larry Bell & IDEEA
Russia President Olga Zakharova

Representatives from several countries at an
IDEEA Federation meeting in Russia

IDEEA USA Board Chair,
Dr. Sylvia Earle

Sylvia after her record 1,250
ft. dive in Molokai Channel

Sylvia in a one atmosphere
articulated hard diving suit

IDEEA USA Board member Will Steger, leader
of a team that walked across Antarctica

Conference venue for IDEEA Two,
The Queen Elizabeth Hotel, Montreal

The Sasakawa International Center for Space Architecture

Students with Buzz Aldrin (2018)

Students with Buzz Aldrin (Circa 1980)

Weightless environments in orbit

Space architectures at SICSA

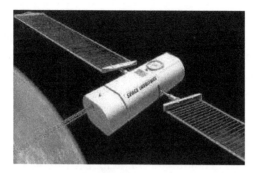

Chapter 8: Pioneering Commercial Space Enterprise

I BELIEVE THAT some of the most exciting and rewarding opportunities for future space development will be realized by entrepreneurs operating in the commercial sector. There is every reason to expect that designers can have important, even leading roles in this arena…just as they can in virtually every other private sector enterprise. And I presume that I can speak with some authority on this subject.

Our Bell and Trotti design and shop was located in a very nice office-warehouse complex that housed another space-related company. Houston-based Space Services Inc. of America (SSIA), headed by enormously respected former Mercury astronaut Deke Slayton, had what many regarded to be an audacious idea that a private entity could infringe upon an exclusive domain of governments…that of launching rockets into space.

On September 9, 1982, they not only believed it could be done, they actually accomplished it. Their privately-funded Conestoga I rocket delivered a 1,000-pound dummy payload (mostly water) to an altitude of 194 miles…more than three times higher than needed to qualify as reaching "space."

SSIA began their business with a bold notion that private enterprise could lower the cost of space launches by clustering a variable number of small engines around a relatively inexpensive expendable booster as necessary to accommodate specific launch payload requirements. The 1982 launch used surplus engines SSIA acquired from the second stage of Minuteman missiles. A later design used surplus 1960's-vintage Scout missile engines.

Back in the mid-1980's NASA was pushing hard to gain public and congressional support for its proposed Space Station Freedom by promoting economic benefits it could serve. NASA had already identified processing of exotic high-value materials in weightless conditions as the prime opportunity. Accordingly, NASA contracted some major management and technology consulting companies, including Arthur D. Little, to investigate and determine which process technologies and products were most promising and to estimate their respective market sizes.

As anticipated, the space market consultants were bringing back exactly the exciting and subsequently publicized revenue projections NASA essentially instructed them to find. All of this suggested that a new era of space commerce might very well be dawning.

A remarkable professional and life adventure precipitated by coincidental contacts with SSIA people combined with our BTI space station work. The opportunity grew out of a chance conversation, as I have since learned, many things do.

Just to keep my personal life record straight here, Cindy and I had amicably divorced by this time. I had become remarried to Nancy, my beloved wife of 37 years and mother of our two sons, Aaron and Ian.

Nancy and I were having dinner with Gui and his then-wife Pampa at a local restaurant when the idea for creating a private commercial material processing space laboratory first came to mind. When I expressed that thought, Gui's eyes opened wide in astonishment. His response left no doubt that a Eureka moment had occurred for both of us.

As a follow-up to our conversation, I contacted every company that had been listed by NASA as a prospective space station customer, and always managed to get in touch with the actual individual who had responded to the survey. While many reposted less interest than was represented, a few were much more encouraging.

Gui had a friend named James Calaway who had just graduated from Oxford. He was very interested in space, and his family was well connected in Houston's top business investment circles. We both met with James, and he was immediately excited about our proposal.

I then contacted Al Louviere, who was then heading NASA's space station planning at the Johnson Space Center, to discuss the idea of him joining us in starting a company. During our lunch meeting he told me that he agreed that the idea was terribly exciting and would love to become involved, but that family responsibilities would make it impossible to risk the financial uncertainty. Alternatively, he suggested that I present the opportunity to Max Faget, who had recently retired as the NASA Johnson Space Center's Chief Engineer.

When I called Max, whom I had never met, asking if we might meet, he inquired how much time I wished to schedule. I replied that twenty minutes would probably be adequate for what I wished to discuss. When I asked if we should meet in his office near NASA, he offered to visit me at SICSA instead.

The planned twenty-minute discussion lasted several hours. When it ended, Max said "I'm with you." What followed, as they say, became the subject of national news features, and left a historic aerospace legacy.

Before I continue any farther with this story, I will share some background about this marvelous aerospace innovator, influential leader, inspirational colleague, and unforgettable friend.

Larry Bell

Max Faget: A Genius Behind Apollo

My life has been honored and enriched to know many wonderful space technology pioneers, along with the astronauts whose lives and missions depended upon them. One, whose design contributions made American Moon landings along with other historic space developments possible, personifies the best of them.

Maxime "Max" Faget, who served as the first Chief Engineer at NASA's Johnson Spaceflight Center, was a technical genius who developed many of the innovative concepts that were incorporated into all manned U.S. spacecraft. Included are the characteristic gum drop capsules for projects Mercury, Gemini, Apollo and the straight-winged design of the Space Shuttle.

Born in British Honduras on August 26, 1921, Max trained in mechanical engineering at San Francisco Junior College and Louisiana State University before doing submarine duty off the coast of Vietnam during World War II. In 1946, following naval service, he was hired by the Pilotless Aircraft Division at the National Advisory Committee for Aeronautics (NACA) Laboratory at Langley Field, Virginia which later evolved to become NASA. NACA's head, Robert Gilruth, later became the first director of NASA Johnson Space Center in Houston.

Max's early NACA work involved solving aerodynamic problems to enable aircraft to fly faster and higher and led to the design of the hypersonic X-15, which achieved Mach 6. These activities later shifted to the design of ballistic missiles.

A June 1997 interview with Max for a *NASA Johnson Space Center Oral History Project* provides direct quotations that bring back wonderful memories of his voice and times.

Commenting about surprise within NACA while planning Project Mercury when the Soviet Union launched Yuri

Gagarin, the first human in space, Max recalled:

> *I think higher management was probably aware of the fact that the Russians were making progress...but it never trickled down to the level of the troops in the trenches, that's for sure.*

Max continued:

> *Nevertheless, we were able to get our first flight off with [Alan] Shepard just a matter of weeks after Yuri Gagarin went into orbit, and I really think that timing made it possible for the President [Kennedy] to jump on the fact that we were in a race with the Russians and that he wanted to win.*

As the NASA Johnson Space Center's top engineer, Max carried the one-man Mercury project through to a two-person Gemini and three-person Apollo capsule along with a two-vehicle Lunar Module landing and ascent system. Gemini was critical to develop and test sophisticated orbital rendezvous and docking and extravehicular activity capabilities which made Apollo possible.

President Kennedy's commitment to extend human presence to the Moon within a decade presented endless scientific and engineering challenges which were both daunting and exciting.

As Max explained:

> *We didn't know what kind of Moon we were going to land on. We didn't know what the radiation environment would be like near the Moon. Just a whole host of things like that we*

didn't really know, and we had to move ahead anyway.

Around the end of the 1960s and beginning of the 1970s, Max led a small group at the NASA Johnson Space Center in planning for an orbital laboratory where people could work to understand microgravity effects both on people and physical processes. That activity led to the development of the Space Shuttle Program.

Max was asked during the NASA interview why go to space at all—why the spark—"do we go for practical business reasons, or is there something more important lurking there in your mind?"

He responded:

> *Man's curiosity is the big driver behind all things that happen, behind all progress... there's always going to be people who have curiosity, and they have to be supported. The same way we've supported the arts, we've got to support the curious.*

Max also emphasized the importance of maintaining a playful spirit:

> *I look at my life, and the way I approach things, everything has been a toy with me. My toys were things that worked, things that flew, dove under the water, little race cars. I always liked things like that, and it was just a hell of a lot of fun to make these things work. And then I grew up and my toys got bigger, more interesting, and I still like to play with toys. So I think the world will always have men that never grow up, and that will do*

> *things that didn't seem to have a hell of a good*
> *purpose at the beginning, but turn out to be*
> *innovative and useful, for reasons that no one*
> *ever dreamed of. So that's the way it goes.*

To me, Max Faget's wisdom and creative spirit will always speak as the smartest technical voice in any room.

The Industrial Space Facility

In 1984, Max, Gui, James Calaway and I formed Space Industries Inc. (SII) to develop and operate a privately-financed orbiting space station called the Industrial Space Facility (ISF). Max became SII's president and CEO, and I was his senior vice president.

The ISF's purpose was to conduct research and produce very high value materials under weightless conditions, Material Processing in Space or MPS as it was termed, represented an important application for NASA's Space Station Freedom that was being proposed to the U.S. Congress at that time.

Although the ISF was truly an independently-orbiting space station, it was only to be inhabited by astronauts in a shirt-sleeve environment when a Space Shuttle was docked with it.

As Max later described this in his 1997 NASA Oral History Project interview:

> *It had an internal volume that was kept*
> *pressurized. Man [or woman] can enter it.*
> *They would live off the life support system on*
> *the Shuttle simply by transferring air between*
> *it and the Shuttle, so we didn't have to put a*
> *lot of life support in there, and when you*
> *wanted to make it bigger, you just add*
> *another unit to it.*

After two or three launches, each unit would be independent of the other. One of the units could be equipped with a life support system to clean the air for all and provide a safe refuge in case of a pressure failure for all of the others. Max explained:

> *...you could get up to maybe six or eight of these things all attached together, and you'd have the equivalent to a Space Station. You would do it where, after the first launch, you're being productive, so you would grow at the same time you would produce. It's a great idea.*

NASA's Space Station Freedom studies were underway at the time we established SII, and our Bell and Trotti firm was actively supporting Boeing. Max had invited Boeing to participate with SII on the ISF venture and Boeing was clearly interested. When asked by Boeing's top management whether I thought that potential involvement by them in the Space Industries ISF venture would harm or benefit their relationship with NASA, I told them I believe it would complement and support their NASA space station work. Boeing agreed, and became a SII business partner. Westinghouse joined the venture in 1978.

ISF's technical concept took shape under Max's leadership, ably supported by his enormously innovative former NASA design colleague Cadwell "CC" Johnson. I prepared a Blue Book describing the overall design and business plan, and private investment solicitations were underway. Those efforts over two years yielded $38 million which supported a small engineering and administrative staff.

Max brought in strong advisory and advocacy support from NASA which greatly enhanced SII's technical and business credibility. Over time, our Board members included

Apollo astronaut Neil Armstrong; the first NASA Johnson Space Center Director Dr. Robert Gilruth; former Tenneco President and CEO, Wilton Scott; American Airlines executive, Charlie Simons; NASA Apollo Mission Controller Christopher "Chris" Kraft, Jr.; 5-mission Shuttle astronaut Joe Allen; and other notables.

Joe Allen succeeded Max as President and CEO, and later became SII's Board Chairman.

The Space Industries, International Saga

In October of 1987, NASA clarified its support for the SII plan in a public memorandum that stated that its role was to "facilitate the development and operations of the ISF." Max then joined with other corporate aerospace leaders in briefing the U.S. Congress on the importance of the space station and shuttle programs in opening up vast commercial opportunities that would benefit the U.S. economy.

It was becoming apparent, however, that NASA was starting to view the ISF as a competitive threat to their space station funding. The White House and Congress were concerned that the budget for NASA's plan was growing at an alarming rate, and many saw private space development as a promising way to trim costs. Some members of Congress even threatened to postpone space station funding until such time that NASA expanded commercial MPS support.

Meanwhile, SII's ISF program continued to gain popular media attention, and some of it had terrible consequences. *The New York Times* featured the program in a front-page article in January 1988, comparing the benefits of the ISF to NASA's space station plan. The real message of the article raised questions about why the government should fund a space station when the private sector could better accomplish its main purpose for a cheaper price.

This event clearly marked the end of any romance

between NASA and SII.

As criticism of the space station increased, some prominent government leaders advocated budgetary cutbacks, and Forbes Magazine published a long editorial suggesting that "it's time to bust up NASA" altogether.

There was little hope of salvaging a cooperative spirit on the part of NASA towards SII as public interest and support for commercial versus government programs continued to grow. Any argument on our part that the ISF would complement and advance the space station's applications for MPS had obviously become futile.

As Max later noted:

> *...and if this [ISF] thing ever got up there and we started adding onto it, people would say, 'Why do we need the Space Station? Why don't we just keep going this way?' It represented a major threat to the continuance of the Space Station Program. It had to be killed, and they did kill it.*

NASA decided to reconsider its earlier agreements with SII, questioning if the exchange of Shuttle launches for lab services might constitute an inappropriate business subsidy. Although the Reagan White House and many members of Congress disagreed with that view, NASA insisted on submitting the matter to some committees for "study." There was no doubt in our minds that this was primarily intended as a delay tactic. And it was successful.

Continued criticism of NASA and its actions towards SII continued to mount. Time magazine lamented a "goodbye to NASA glory days," and Aviation Week, the leading aerospace publication, openly faulted NASA for frustrating commercial space entrepreneurism.

Although the Reagan White House had warned NASA

that if they didn't want the ISF program, it didn't need a space station either, NASA continued to lobby against our plan. In March 1988, they won and we lost by a one-vote margin in Congress.

Business Week highlighted the NASA-ISF controversy in a feature about "a space station that's losing its boosters," and *Air and Space Magazine* presented a comprehensive story about politics surrounding the SII matter that is probably the best and most complete explanation ever published on the subject.

With hope for our ISF initiative waning, SII turned to other revenue opportunities. The company began to design and build space hardware for government, corporate and academic organizations. One of its projects was a Wake Shield experiment system that was developed for the U of H's Space Vacuum Epitaxy Center (SVEC). The device flew on two Space Shuttle missions.

The new fee-for-service business opportunities continued to finance ISF planning and engineering design as our company endeavored to keep the original dream alive. Customers included the Russia Government, as SII created equipment systems for its Mir Space Station.

By 1990, SII had become the 3rd fastest growing company in Houston according to a Houston Post business survey. As noted in the headlines of an *Industry Week* article, SII's "profits tempered disappointments."

Recognition of SII's survival achievement and international business successes under the new leadership of President Joe Allen and CFO David Langstaff was evident in a long article that appeared in the *New Yorker Magazine.* By 1992, the firm was producing more than $10 million in annual revenues, making it an attractive high-tech merger candidate.

In 1993, SII merged with Calspan, a much larger company with more than 3,000 people. Calspan acquired one-third ownership and took over management of SII. Calspan is a leading test lab for advanced aircraft used by all U.S. military agencies and for automotive crash testing in Detroit.

Joe Allen had served on the Board of Directors of Calspan and continued on as Calspan's president following the SII merger that created Calspan SRL.

These enterprises led, in turn, to a merger with Veridian, a national leader in providing information security services for government and industry. The company went public on the New York Stock Exchange in 2002 and was the second most successful Initial Public Offering (IPO) in the United States that year. Having grown to employ more than 8,000 professionals, the corporation was purchased by General Dynamics for about $1.5 billion.

Entirely by coincidence, I had also served with Max on the board of another company, Winzen International, which developed enormously large high-altitude balloons used for edge-of space weather recordings. In fact, we were the only outside board members at that time.

Max and I flew to Dallas, then rented a car for the 80-mile drive northeast to Sculpture Springs, Texas where Winzen International was located. There, workers in narrow, 300-foot-long buildings carefully hand fabricated and developed huge plastic balloons to carry scientific instruments for NASA and other clients to stratospheric altitudes out of fragile polyethylene plastic material thinner than a human hair. When inflated they might contain as much as 10 million cubic feet of hydrogen or helium.

Winzen's on-site facility also manufactured the polyethylene material which demanded compliance with very exacting quality control standards.

I particularly enjoyed attending the annual employee award events when justifiably proud employees and management gathered together on folding chairs set up on the lawn. I can still imagine one of those wholesome scenes depicted in a Norman Rockwell painting.

Sadly, after weather satellites came along, there simply was no longer a large enough market to sustain Winzen's business. We had no alternative but sell the balloon operation to Raven Aerostar, the only remaining competitor.

More ironically, that high-tech, stringently quality-controlled, thin-film polyethylene production plant was sold as well. It now produces packaging material for Tyson Foods.

Max Faget finally left this world in 2004. The honor of working with him and enjoying time in his wise company over a period of two decades has been a treasured gift in my life.

A New Space Enterprise Adventure

As with SII, another commercial space firm that began as a chance conversation is International Space Enterprises (ISE), the company now producing hybrid-electric vehicles. During a coffee break at a space conference in Albuquerque I was joined by two old friends that I introduced to each other. Mike Simon was a Stanford University aerospace engineering graduate who was working at General Dynamics, and George Schuh was a scientist at Sandia National Laboratories.

Our discussion turned to the subject of private space enterprise, and I suggested that new opportunities might be afforded by purchasing rockets and other assets from Russia. By the time we left the table some hours later we had agreed to organize a venture to act upon that possibility. ISE had just been born.

Soon afterwards, we invited two others to join us in creating the company. I had known Valery Aksamentov as a fellow teacher at summer sessions organized by the International Space University in various countries. David Mazaika was a professional colleague of Mike Simon at General Dynamics and headed its business development activities.

Our founding goals were spectacularly ambitious, particularly for a new company with no financing:

- To develop and launch inexpensive tele-operated rovers that would enable young students to explore the Moon from their classrooms.

- To use the moon as a platform for deep space astronomy applying advanced ultraviolet telescopes.

- To place scientific equipment on the Moon to collect and analyze surface samples and search for water.

- To install instruments and sensors to study lunar geophysics and the formation of our Solar System.

- To apply advanced technology to provide virtual exploration experiences for "armchair astronauts" on Earth.

To briefly summarize a long and complicated story, we proposed a joint venture plan to top executives of some of Russia's leading space organizations. More specifically, we offered to raise necessary money to purchase and launch a dozen of their huge Proton rockets at about $20 million each. These vehicles are capable of placing a metric ton of payload on the Moon. We also offered to pay Russia to upgrade their lunar landers to accommodate payloads of that size, and to purchase some of their rovers for our purposes.

Our plan provided that we would share our profits from scientific customers with them. Following a series of meetings in Moscow, they approved our proposal, and we formed a joint venture named ISELA. That term that combined the initials of ISE and the first letters of our primary partners, Lavochkin. Roughly translated, ISELA also means "and landed" in Russian, which seemed very appropriate for our lunar surface program.

We later met with NASA's administrator, Dan Goldin, to offer NASA an opportunity to participate in our scientific plans. NASA was not enamored with the idea of flying experiments on Russian rockets at that time, and they also controlled most scientific markets we were seeking. As a result, our primary missions were grounded.

However, our efforts were not without some achievements. We managed to raise some funding from aerospace companies to pay for upgrades to the Russian lander design and to work with Lavochkin to develop some advanced rovers. ISE then began to design and build a variety of its own rovers for government and private applications, including some that could be used at hazardous toxic sites and for explosive device retrieval operations.

And while we never launched anything on the Moon ourselves, we had an unexpected role in another initiative that did. When we originally presented our plan to NASA, we introduced the administrator to the representative of another small company that was developing a satellite that could be placed in lunar orbit to search for traces of water.

We proposed that there was ample excess capacity on a Russian Proton rocket to carry that experiment along with ours. That meeting led to NASA's funding of the Lunar Prospector Mission which flew on a U.S. rocket and provided evidence that water does exist on the Moon.

You just never know what a conversation over coffee can ultimately lead to.

Orbital Advertising Goes Prime Time

What, exactly, does space advertising have to do with design? To be honest, I really don't know, other than it lies along another unexpected coffee cup-caused design trajectory that has delivered me to the edge of space, as I will soon explain.

My space advertising experiences began at International Space Enterprises during the early 90's...the company that wanted to launch experiments to the Moon, and instead, launched a hybrid-electric vehicle business.

Larry Bell

Valery Aksamentov made arrangements to fly a four-foot-tall Pepsi can mockup that was photographed from outside the Russian Mir Space Station for a Super Bowl commercial. Interestingly, it was an American astronaut, Shannon Lucid, who took the picture.

Valery and I co-founded another company called Globus Space in 1995. Its primary purpose was to arrange commercial advertising programs from other U.S. corporations that would provide badly needed revenues for Russian space organizations.

In 1999, we undertook an assignment to place a thirty-foot-tall Pizza Hut logo on the Russian Proton rocket that launched the first module to the International Space Station from Baikonur, Kazakhstan. The historic launch of the Russian Service Module occurred as planned on July 13, 2000.

Baikonur, the location of all Russian Launches, is not an easy place to get to. During the Cold War it served as an ICBM site, and we can imagine that many of those missiles were programmed to fly in our direction. It is a huge area, covering hundreds of square miles of desert in Kazakhstan.

Since no commercial airlines fly to the area, a small group of us leased a YAK-40 aircraft complete with crew for the four-hour flight from Moscow to view the launch. Security was very tight at the site, and all attendees required official invitations from the Russian Government along with Kazakhstan visas.

We received very special treatment during our trip. With world attention focused on the placement of the initial element of the International Space Station in orbit, it seemed incredible that we were the only visitors allowed at the site on the day before that historic launch. Even NASA representatives were denied that opportunity.

The following day the launch went off on schedule. When I asked a Russian official why there was no customary count-down that is witnessed at NASA launches, he simply shrugged his shoulders and offers a simple reply: "When it is supposed to go, it goes." He was right.

Part of Pizza Hut's promotional program required arrangements to film a cosmonaut eating their product in space while wearing their logo on his cap. This wasn't as simple as you might imagine. All food had to be officially checked out and approved before it could be launched. Accordingly, the pizza was actually made from a special authorized recipe in Russia. Fortunately for us, Pizza Hut and the cosmonauts, no indigestion occurred.

Now I will share my near-space crash and burn experience.

The next Proton launch delivered the first crew to the International Space Station. Although I was not there in person, at least my name was. It was placed along with the names of four of my partners in the 1st stage of the rocket by the Russians.

As a result, I accompanied those astronauts and cosmonauts in name and spirit to the edge of space. That was before I then crashed, as planned, to join the camels on the surface of Kazakhstan's desert. That put an ignominious end to my astronautic career.

Sadly, no tickertape parades for me.

Hotels with Spectacular Views

Early one Monday morning, I received a phone call at SICSA that I thought was a joke. The caller introduced himself as Robert Bigelow, a billionaire who owned a large chain of Budget Suite hotels and who wanted to build one in space. He said he wanted some help in designing it.

Before I rejected the call as a prank, he said he had been referred to me by Charlie Walker, an astronaut that I know quite well. Charlie had a good sense of humor, but I didn't think that he would go that far using it.

I suggested that Mr. Bigelow visit SICSA and a few weeks later he did, private twin-engine jet and all. As it turned out, he was exactly what he was presenting himself to be.

Robert Bigelow is a remarkable man who pursues his dream. He had built a large facility in Las Vegas where his company, Bigelow Aerospace, developed and tested inflatable modules for space tourism applications. These modular soft-walled structures are compactly folded around metal cores for launch and deployed by inflation upon reaching orbit.

The orbital hotel we designed for Robert Bigelow is somewhat different from the one he is currently developing. Ours was designed to accommodate up to 100 people not including a small support crew. It included inflatable Space Media Lab and Habitat Modules, along with smaller conventional hard modules for medical and other accommodations.

Bigelow Aerospace launched two small-scale expandable module test prototypes to low Earth orbit aboard Russian rockets. The first, Genesis, was deployed in July 2006, followed by Genesis II in June 2007.

The company also deployed a much larger 565 cubic foot prototype, the Bigelow Expandable Activity Module (BEAM), aboard Elon Musk's SpaceX Falcon 9 rocket which is attached to the International Space Station. BEAM is being tested to determine how well it holds pressure and provides protection from space radiation.

Robert Bigelow's goal was to market individual and clustered commercial space habitats of substantial size. A planned BA 330 series (referring to 330 cubic meter/ 11,700 cubic foot volume) would be about 45 feet long, 22 feet in diameter, and weigh about 43,000 pounds.

An even larger 2,100 cubic meter Olympus, contemplated for future use in Earth's orbit and beyond would have nearly twice the pressurized volume capacity available with the entire International Space Station. With a deployed diameter of approximately 41 feet and weight of 79-90 tons, it will require a super-heavy launch vehicle for delivery to orbit.

Be warned, however, Bigelow's terrestrial Budget Suite room rents won't apply. A pricing scheme posted on the Bigelow Aerospace website in 2015 suggested that a future $51.25 million rental fee will enable a customer to experience 60 days aboard a BA 330, including all transport, training, and food. As for the good news, it advertises: "Bring your clothes and your money. We provide everything else."

Max Faget

Joe Allen

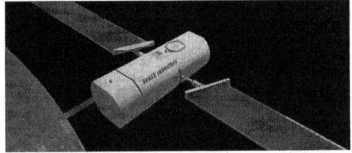
Space Industries International, the Industrial Space Facility (ISF)

Material processing

Neil Armstrong

The Russian rocket Pizza Hut launch from Kazakhstan

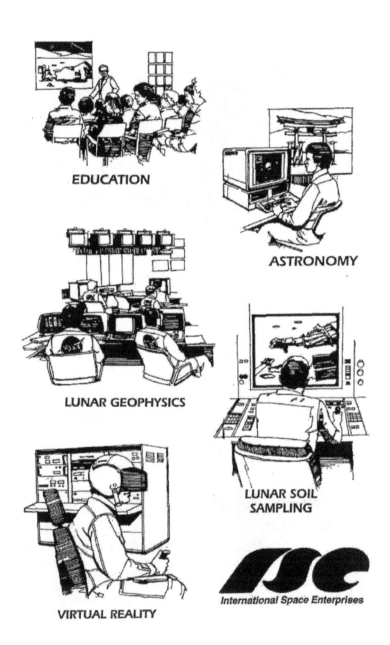

EDUCATION

ASTRONOMY

LUNAR GEOPHYSICS

LUNAR SOIL
SAMPLING

VIRTUAL REALITY

International Space Enterprises

US-Russian lunar lander design

International Space Enterprises (ISE)

Chapter 9: New Space Design Frontiers

SPACE EXPLORATION ISN'T only about astronautics. Some of the most important pioneers may be those who plan and provide the business incentives and technical innovations that make space development economically viable and sustainable.

A new breed of commercial space entrepreneurs envision future businesses leading beyond low Earth orbit to surfaces of the Moon and Mars. Robert Bigelow has predicted:

> *I think expandable systems hold the key...our long-term goal as a company is to have a lunar base that might be a modest size, initially. In somewhere around 2023.*[16]

Former PayPal CEO Elon Musk, founder of SpaceX, has said that his company was prepared to send colonists to Mars by 2024, and that a self-sustaining city on Mars could be achieved

---

[16] Adam Higginbotham, *Robert Bigelow Plans a Real Estate Empire in Space: A Las Vegas Billionaire Plans to Build a Real Estate Empire in Space,* May 2, 2013, Bloomberg.com, http://www.bloomberg.com/news/articles/2013-05-02/robert-bigelow-plans-a-real-estate-empire-in-space

in 40 to 100 years.[17]

Given my background, it should come as no real surprise that many of our SICSA studies investigate ways that the private sector can play vital roles in advancing commercial lunar-planetary space development in combination with government-sponsored programs. Much of this work explores means to minimize mission costs, maximize long-term crew sustainability and grow evolutionary operational capabilities using surface resources.

Living off the Land

Space affords a natural supply depot stocked with a vast assortment and quantity of potentially useful, even critical, materials that can expand human experience and enterprise.

Of these rich resource caches, the Moon is the closest, representing a relatively near-term source and laboratory for rocket propellant, oxygen and water production, and an operational base for development and demonstration of other extraterrestrial technologies.

The Moon has some similarities to Earth, relative close access to solar energy, some gravity and plentiful resources such as water, oxygen and minerals. However, there are also many dramatic differences. Included are the lack of any atmosphere, extreme temperatures and the fact that harvesting and processing those resources to support substantial scale operations will impose daunting technological and infrastructure challenges.

Recent discoveries of large quantities of water on the Moon have excited great interest for a multitude of

[17] *Elon Musk: I'll Put a Man on Mars in 10 Years*, April 4, 2011, Market Watch (New York: The Wall Street Journal,
http://www.wsj.com/video/elon-musk-ill-put-a-man-on-mars-in-10-years/CCF1FC62-BB0D-4561-938C-DF0DEFAD15BA.html

applications which can greatly reduce dependence upon costly transportation of Earth-delivered consumables. The precious molecule is vital to life for drinking and as a constituent source of oxygen for breathing. It is also a crucial source of hydrogen and oxygen for rocket propellant to fuel ascent vehicles from the lunar surface and to cis-lunar (Moon-Earth-orbit) space and beyond.

Accessing, separating and processing those substances, on the other hand, will present enormous challenges. In addition to the fact that enormous costs of delivering cargo to the Moon will stringently limit equipment sizes, inventories and maintenance spares, harsh and unique environmental conditions will impose a variety of operational difficulties. All such operations and maintenance will need to be highly automated with little or no human hands-on intervention when something inevitably goes wrong.

Airless vacuum and low gravity ($1/6^{th}$ Earth gravity) conditions will impact many operations and systems related to digging, separating, processing, containing and transporting resource materials. In addition, huge temperatures variations ranging from about $-275°$ F (night) to about $+280°$ F (day) and differences between sunlit and shadowed areas will reduce the viscosity of equipment lubricants and degrade battery efficiencies.

Destination Mars

While the Moon's relatively close proximity to Earth and diverse surface resources offer special advantages, the more Earth-like characteristics and greater resource variety afforded by Mars make it far more attractive as a true human destination.

Of special importance, recent findings obtained from NASA's Mars Reconnaissance Orbiter (MRO) offer evidence that as with the Moon, water also exists on Mars, and that perhaps some even intermittently flows on its surface.

Mars also has a great diversity of other resources, some of which are present but most likely less accessible on the Moon. For example, carbon dioxide, nitrogen and hydrogen exist on the Moon, but unlike Mars, only in tiny parts per million quantities. While oxygen is abundant in lunar soil, it is tightly-bound in oxide which requires high energy processes to release.

Like Earth, with an atmosphere and an axial tilt (25.2°), Mars has seasons and weather which varies somewhat from year-to-year. Although Mars days are nearly the same length of time as on Earth (24 hours and 37 minutes), each Mars year equals 687 Earth days, about two Earth Years.

Mars is about 50% farther from the Sun than the Earth and Moon are, and also different. The airless Moon doesn't filter away surface sunlight as the planetary atmospheres do. Massive planet-wide dust storms on Mars present an extreme case. Yet while on one hand these features afford some lunar surface solar energy benefits, the Moon's 28-Earth-day light-dark cycle counteracts these advantages.

Like Earth, Mars offers true climate variety. Surface temperatures average about -67°F, can reach a high of +68°F at the equator (noon), and drop to a low of -243°F at the poles. NASA's Spirit rover recorded a maximum daytime air temperature in the shade of 95°F, and regularly recorded temperatures well above 32°F except during winter.

Beyond Footprints and Flagpoles

My very close long-time friend and colleague, Gemini 12 and Apollo 11 Astronaut Buzz Aldrin, has a plan to efficiently and continuously cycle future space pioneers and settlers back and forth between new worlds upon orbital expressways. These Aldrin Cyclers will apply celestial mechanics—Newton's laws of motion—using gravitational forces of the Earth, Moon and

Mars to minimize trip times and fuel costs.

The Cycler idea is somewhat analogous to cruise ships that drop off and take on passengers without pulling into harbor, except that Aldrin Cyclers don't stop when they fly by Earth. Passengers access the spacecraft via speedy space taxis that catch up with them.

Like catching a bus that repeats the same route over and over, one has to run fast to catch it. In this case, one must travel very, very fast to catch that space bus. Intercepting it from Earth requires a rendezvous velocity of about 6 kilometers per second (13,400 mph).

Using two Cyclers, the frequency of Earth-Mars encounters for back and forth crew and cargo transfer opportunities is about 2 1/7 years. Once onboard, a passenger's time-of-flight from the vicinity of Earth to Mars would take about six months.

Some Special Buzz Words

Buzz strongly believes that many of those future Mars travelers will be settlers rather than tourists. As he emphasizes in our recently co-authored book, *Beyond Flagpoles and Footprints:*

> *This is far different than Apollo expeditions to the Moon where voyagers do some experiments, plant a flag and claim success. Having them go to the Red Planet and repeat this, in my view, is senseless. Since great distance between Mars and Earth makes a feasible return window very narrow, it makes far more sense to transport people there who plan to stay. So yes, I suggest that going to Mars means preparing for permanence on the planet.*

Permanent Mars settlers will be 21st Century pilgrims, pioneering a new way of life. That will indeed take a special kind of person.

Instead of the traditional pilot/scientist/engineer, Martian homesteaders will be selected more for their personalities...flexible, inventive and determined in the face of unpredictability.

As Buzz recognizes:

> *Inevitably, this will invoke daunting risks and casualties, just as other pioneering ventures have. Unfortunately, pioneers will always pave the way with sacrifices. Over the decades, we have lost numbers of individuals—several of them close personal friends of mine—all intent on pushing the boundaries of exploration and seeking new horizons. Risk and reward is the weighing scale of exploring and taming space.*

Buzz teaches by his own example that if you want to do something significant, something noble, something that perhaps has never been done before, you must be willing to fail. And don't be surprised or devastated if you do. Untold numbers of people have experienced major failures and have recovered to become not only successful, but also better, stronger people.

As he puts it:

> *Failure is not a sign of weakness. It is evidence that you are alive and accepting of worthwhile risks.*

Landing habitat module on Mars

Prospector spacecraft on the Martian moon Phobos

Mobile Mars habitat module

Mars surface habitat with inflatable module

Chapter 10: Thinking Beyond Boxes and Boundaries

I BELIEVE THAT true innovators and doers must be willing to reach beyond closed mind boxes and rigid conceptual boundaries others take for granted in order to remodel realities to match imagined possibilities. In doing so, they convert what-ifs, to what-is's.

Innovators and doers retain child-like optimism. British celebrity chef Heston Blumenthal observed:

> *As we get older, we tend to become more risk averse because we find reasons why things won't work. When you are a kid you think everything is possible, and I think with creativity it is important to keep that naivety.*

Visualizing unbounded possibilities may be a forgotten art we must relearn from our inner child-selves in order to rediscover what we truly value most in ourselves and our lives. Those sorts of visions probably came easier in childhood before setbacks and disappointments challenged our confidence, before we forgot how special we are and before magic was disproven.

Lofty purposes and goals should be regarded as more than

idle fantasies. Dreaming up worthwhile futures can also energize us to make them become real. If we don't visualize what we really want out of life, we won't have any basis for setting our course in the right direction in order to avoid wasting time and energy on routes destined to nowhere.

Assuming that visions direct realities, and I believe they do, then dreaming up higher possibilities is something that we can't afford to abandon or postpone. If we do, our ship of opportunity piloted by others may leave port without us.

Michelangelo purportedly warned:

> *The greater danger for the most part of us lies not in setting our aim too high and falling short; but in setting our aim too low, and achieving our mark.*

Six centuries later American media mogul and philanthropist Ted Turner offered similar advice: "Set your goals higher than you can achieve."

Innovators and doers are often too interested in what they are doing and committed to dreams and goals they are pursuing to waste time worrying about failing. In other cases, having assessed risks that things won't work out as planned, they decided that penalties of not trying will be far greater.

Achievers are characteristically motivated, persistent and resilient. They don't quit. They experiment, often fail, learn more and repeatedly start over.

Thomas Edison, as quoted in a 1921 interview by B.C. Forbes for *American Magazine*, said:

> *I never allow myself to become discouraged under any circumstances. I recall that after we had conducted thousands of experiments on a certain project without solving the problem, one of my associates after we had conducted*

> *the crowning experiment and it had proved a failure, expressed discouragement and disgust over our having failed to find out anything. I cheerily assured him that we had learned something. For we had learned for a certainty that the thing couldn't be done that way. And we would have to try some other way. We sometimes learn a lot from our failures if we have put into the effort the best thought and work we are capable of.*

Calvin Coolidge, the 30th President of the United States, a man noted for decisive actions, attributed central importance to dogged determination.

He said:

> *Nothing in this world can take the place of persistence. Talent will not: nothing is more common than the unsuccessful men with talent. Genius will not: unrewarded genius is almost a proverb. Education will not: the world is full of educated derelicts. Persistence and determination alone are omnipotent.*

American engineer, inventor, Delco founder and General Motors research director Charles Kettering advised that true innovators don't take fear of failing very seriously.

He wrote:

> *You see, from the time a person is six years old until he graduates from college he has to take three or four examinations a year. If he flunks once, he is out. But an inventor is almost always failing. He tries and fails maybe a thousand times. If he succeeds once, then*

he's in. These things are diametrically opposite. We often say that the biggest job we have, is to teach a newly hired employee how to fail intelligently. We have to train him to experiment over and over and to keep on trying and failing until he learns what will work.

Transcending Comfort Zones

Thinking "out of the box," whether applied to architectures or other structures of convention, often smacks of unwelcome nonconformity to prevailing and popularly accepted conventions and standards. Doing so often requires breaking some molds, including those we have allowed to crystalize around our own mindsets.

Steve Jobs cautions:

> *Your time is limited, so don't waste it living someone else's life. Don't be trapped by dogma—which is living with the results of other peoples' thinking. Don't let the noise of others' opinions drown out your inner voice. And most important, have the courage to follow your heart and intuition. They somehow already know what you truly want to become. Everything else is secondary.*

Original thinking springs much less a desire to be different than from a willingness to be different in pursuit of something fresh or better. The well-known writer, reporter and political commentator Walter Lippman famously quipped: "When all think alike, then no one is thinking."

Innovative thinking sometimes requires getting outside of our comfort zones of tried-and-true experiences, ready-made methods and popularly-assumed perspectives. Inventor Charles Kettering characterized this as to *"Get off Route 35."*

In her book *Breakthrough Creativity: Achieving Top Performance Using the Eight Creative Talents,* Lynne C. Levesque emphasizes that creative people are unafraid to challenge the status quo:

> *To be creative you have to contribute something different from what you've done before. Your results need not be original to the world; few results truly meet that criterion. In fact, most results are built on the work of others.*

The late American scientist and inventor Edwin H. Land, who co-founded the Polaroid Corporation, once said: "The essential part of creativity is not being afraid to fail."

And as my charismatic High School band director, Irv Hansen, once told us:

> *Don't be afraid to blow a sour note...be prepared to make it a loud one everyone can hear. Otherwise, we will never make music.*

Those who know me will recognize that I continue to heed this wisdom. Yet many years later, when I reminded him of this important advice, he said he didn't remember. Maybe he just didn't want to accept blame on my behalf.

Important life decisions often involve tradeoffs between desires for security and adventure. Since we can't have copious amounts of both, we typically try to hedge our overall bets so that we can expect to wind up with an acceptable balance.

When we have too little security to meet basic needs, it is natural to be risk-averse. When we are comfortable and confident that those requirements are assured, then we can well afford to pursue more adventurous options.

Being too risk-averse, however, may jeopardize personal growth. As Neale Donald Walsch, American author of the series *Conversations with God* wrote: "Life begins at the end of your comfort zone."

English actor Dan Stevens regards the personal "comfort zone" as the great enemy of creativity, He advises: "Moving beyond it necessitates intuition, which in turn configures new perspectives and conquers fears."

In reality, there is no such thing as total security. Regardless how affluent and satisfied we are, mortality presents inescapable perils for everyone. And if we hope to achieve more security in life, we must almost always accept some risks and costs associated with the new initiatives required to gain it.

When we avoid worthwhile risks, we deny ourselves challenges that cause us to grow and to realize the excitement of experiencing fuller lives. Henry David Thoreau reminds us: "The price of anything is the amount of life you exchange for it."

Each of us must determine which risks and costs are appropriate. When opportunities come along, no one else can evaluate whether the prospective benefits warrant necessary expenses and uncertainties of the gamble…whether they will forward off cliffs or open up new pathways of growth.

I have taken some big life-changing risks that I am very grateful for. For example, I left a dead-end at Inland Steel in Milwaukee to experience new learning adventures in the U.S. Air Force, and left a closed future as a full professor with tenure at the University of Illinois to explore more open possibilities in Houston.

In both instances, friends and associates thought I was crazy. Mostly, I was simply trading fixed certainties for more flexible experience and growth potentials. With prospects for a full life in the balance, I instinctively saw settling for less as the greatest risk of all.

Personal and professional growth entails a constantly evolving state of awareness and development...an open-ended pursuit of understanding and a perpetual process of "becoming." Opportunities for progress are stunted when we surrender important priorities and potentials to establishment viewpoints, orthodox ideologies and groupthink expectations.

I credit most major professional opportunities and rewards in my own life to stubborn "maverick" resistance to all of the above. Although such experiences and confrontations have been, and continue to be, unwelcome, discouraging and exasperating, I owe any true value of this book to them.

My earliest open view of architecture as places of activity and movement, rather than more predominately conceived as encapsulating structures, motivated me to innovate Synchroveyor. That exploratory experience introduced me to discover passion in sculpture and joys of industrial design...which somehow incongruously led me to involvement in urban crime prevention and economic development...followed by architectures and commercial business initiatives in space...which loops me back to my Greenland experience and other extreme environments on Earth and beyond... including, most recently, solar and other influences upon out planet's climate.

From Rocket Science to Climate Fictions

Speaking of expanding beyond comfort zones, publishing hundreds of opinion columns in major news publications about controversial topics, climate change and energy in particular, has extended my boundaries into territories of perilous risk.

After all, why would a common rocket scientist presume to know about anything so complex, specialized and sophisticated as climate change and influences on energy policies? Shouldn't that sort of stuff be left to *real* experts? If so, exactly who are they?

One is Dr. Fred Singer, and he's the one who first got me involved in a lot of research, thought and networking with other true experts from around the world on this topic.

Fred was an internationally recognized climate physicist and Distinguished Research Professor at George Mason University. He served as the first director of the U.S. National Weather Satellite Service, and also as vice chairman of the U.S. National Advisory Committee on Oceans and Atmospheres. In addition, he has authored numerous books about climate, energy and environmental issues, including a recent *New York Times* bestseller, *Unstoppable Global Warming*.

I first became acquainted with Fred through common space exploration interests. During a visit to my office he happened to mention that satellite temperature recordings of the Earth's lower atmosphere were cooling more rapidly, relative to the surface, than greenhouse theory predicts. It would be expected that carbon dioxide would warm the lower atmosphere first, which would then radiate heat back to the surface, the reverse of what was being observed.

Given that our primary conversation was focused upon Mars, not Earth, I didn't give the matter much thought until a year or more later when I was contemplating possible lessons that might be applied from the way natural climate operates on Spaceship Earth to the design of artificial life support and energy systems operating beyond our planet. In other words, I began to investigate climate phenomena from holistic references regarding basic principles that govern how natural and technical systems work, how they connect together and how they can be managed to support the most complex systems of all—humans.

Although I came to appreciate that climate is a staggeringly more complex issue than the comparatively simple technical workings of spacecraft systems, it was the very nature of these influences and conditions that led me to question the simplistic and alarmist climate influences attributed to man-made CO_2 emissions by prominent members of the "scientific establishment." This quest for understanding motivated and informed my authorship of two books on the subject.[18, 19]

The second of these books is dedicated to Fred, who along with tens of thousands of other very courageous scientists and writers, have suffered personal attacks and professional penalties for challenging biased research and politically-driven policy agendas. Included are character assassinations, and sometimes even threats upon their personal safety and lives.

Several other distinguished friends and associates with expertise in addressing complex aerospace systems have also become deeply involved in climate and energy science issues. I'll share some observations from a couple of them.

As presented on the back cover of *Climate of Corruption*, Apollo 7 Astronaut Walter Cunningham writes:

> *Those of us fortunate enough to have traveled in space bet our lives on the competence, dedication, and integrity of science and technology professionals who made our missions possible. From space the Earth does appear fragile, but on returning to terra firma, we are reminded of how well our home planet has survived eons of extreme heating and cooling.*
>
> *In the last twenty years, I have watched the high standards of science being violated by*

[18] *Climate of Corruption: Politics and Power Behind the Global Warming Hoax*, (2012).
[19] *Scared Witless: Prophets and Profits of Climate Doom*, (2015).

a few influential climate scientists, including some at NASA, while special interest opportunists have dangerously abused our public trust.

Today's real man-made crisis is an attempt to scare the public into thinking current temperatures are unusual, that humans are responsible for it, and that we can control the temperature of the Earth.

This important book shines light on the self-serving agendas and shoddy political dealings behind the global warming hoax that we absolutely must change while there is still time."

I posted an interview with legendary aerospace designer Burt Rutan regarding his extensive studies on the climate topic in my September 9, 2012 *Forbes* column.

Some excerpts of Burt's comments follow:

[A]bout three or four years ago many alarmist claims by some climate scientists caught my attention. Since this is such an important topic, I began to look into it firsthand.

I do have considerable expertise in processing and presenting data. I have also had extensive opportunities to observe how other people present data and use it to make their points.

The first thing that got my attention, a lot of people's attention, was statements that the entire planet is heading towards a future climate catastrophe that is attributable to human carbon dioxide emissions. It's obviously an extremely important issue which has gotten a huge amount of media attention. I was particularly concerned because the proposed solutions will have enormous impacts upon costs of energy, which of course, will increase costs of everything.

I was struck by claims that we are experiencing unprecedented warming caused by Man, where data clearly shows that our recent warming isn't unprecedented.

Actually, looking back over the past 11,000 or so years since Earth began to recover from the last big Ice Age, we're experiencing a very moderate and stable climate stage. And going back nearly half of the past million years, a long Ice Age occurred about every 90,000 years or so with a large percentage of the planet uninhabitable.

Have humans had any influence on climate? Sure, probably so, although no one has ever succeeded in accurately measuring them. In the absence of everything else, would adding carbon dioxide to the atmosphere have produced some warming? Again, yes. Answering these two questions, and these two questions only, you will see a very large consensus, not only among alarmists, but essentially, every skeptic would also agree.

But none of this presumed warming should be taken to suggest that the results will be catastrophic, causing terribly dangerous things to happen... like serious heat waves and droughts which cause crops to fail... or that when they occur they are "unprecedented." It doesn't require anyone with a climate science-

related degree to recognize, for example, that
1938 was the warmest year in recent times,
and that CO_2 levels were much lower then.

Expanding Our Personal Worlds

We all know people who sell large portions of their lives in exchange for doing things that they don't enjoy, and sometimes even detest. We hear them tell us that they are overworked, underpaid and unappreciated; their job is boring, and their colleagues are offensive. Or they feel alone and neglected when others who they haven't taken initiative to reach out to haven't called or visited them.

Sadly, some people who feel trapped in unsatisfying lives very well may have few escape options and are left to make the best of bad circumstances. For example, when other people who depend upon them would make external employment changes or relocations impossible. A primary option may be to innovate ways to change their lives from inside their minds.

As British-born neurologist and science writer Oliver Sacks instructed in a 2012 *New Yorker* magazine article:

> *To live on a day-to-day basis is insufficient for*
> *human beings; we need to transcend,*
> *transport, escape; we need meaning,*
> *understanding, and explanation; we need to*
> *see over-all patterns in our lives. We need*
> *hope, the sense of a future. And we need*
> *freedom (or, at least, the illusion of freedom)*
> *to get beyond ourselves, whether with*
> *telescopes and microscopes and our ever-*
> *burgeoning technology, or in states of mind*
> *that allow us to travel to other worlds, to rise*

above our immediate surroundings.[20]

Like all living organisms, when we cease to progress, we get into big trouble. Either we grow or stagnate—those are our only options. But unlike other less cerebral and self-actualizing creatures, we can exercise free will to determine which condition shall prevail. We can elect to be vital parts of the happenings around us—or we can passively submit to the forces of those events, whether they sweep us along or pass us by.

In short, we can become involved, or we can become obsolete.

Few of us would consciously choose that second option. It is more fun to be productive and witness wonderful changes that we can bring about through our enterprises. It's gratifying to engage and cooperatively interact with people that we respect and enjoy. It's rewarding to realize that we are in control of our lives, and not allow ourselves to be dominated by circumstances that limit and misuse us.

It's important to remember that age is a self-fulfilling attitude. We don't really grow old. We become old when we allow ourselves to stop growing.

Growth and vitality, versus stagnancy and obsolescence, are matters for personal definition. Each of us must determine which progress indicators are appropriate. Some may attach such priorities as wealth and material possessions, popularity and recognition, work achievements and honors, activity levels and adventures, spiritual and intellectual understanding, compassion and service, qualities of relationships and triumphs over tragedies and hardships.

Writing in The New Yorker, David Brooks stresses the

---

[20] *Altered States: Self-experiments in Chemistry*, Oliver Sacks, The New Yorker, August 2012.

importance of connections between our lives and others:

> *I've come to think that flourishing consists of putting yourself in situations in which you lose self-consciousness and become fused with other people, experiences, or tasks. It happens sometimes when you are lost in a hard challenge, or when an artist or a craftsman becomes one with the brush or the tool. It happens sometimes while you're playing sports, or listening to music or lost in a story, or to some people when they feel enveloped by God's love. And it happens most when we connect with other people.*

Brooks concludes:

> *I've come to think that happiness isn't really produced by conscious accomplishments. Happiness is a measure of how thickly the unconscious parts of our minds are intertwined with other people and with activities. Happiness is determined by how much information and affection flows through us covertly every day and year.*[21]

Most of us apply multiple criteria. The central issue has to do with how we visualize ourselves today relative to where we were and where we are heading.

---

[21] *Social Animal*, David Brooks, The New Yorker, Annals of Psychology, January 17, 2011.

Climate scientist Dr. Fred Singer   Apollo astronaut Walt Cunningham

Burt Rutan (L) with Max Faget and Buzz Aldrin

Buzz, Neil and me

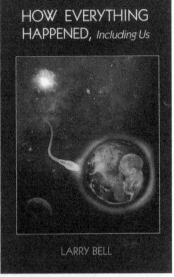

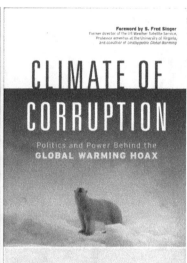

Chapter 11: Innovating Fuller Lives

PSYCHOLOGIST/WRITER EDWARD DE BONO advises that: "We need creativity in order to break free of temporary structures that have been set up by a particular sequence of experience."

De Bono observed:

> *Creativity is a great motivator because it makes people interested in what they are doing. Creativity gives the possibility of some sort of achievement to everyone. Creativity makes life more fun and interesting.*

De Bono believes: "Creativity involves breaking out of established patterns in order to look at things in a different way."

He counsels that,

> *One very important aspect of motivation is the willingness to stop and look at things no one else bothered to look at. This simple process of focusing on things that are taken for granted is a powerful source of creativity.*

Hungarian biochemist Albert von Szent-Gyorgyi, who discovered Vitamin C and who won the Nobel Prize in 1937, characterized creative imagining as recognizing possibilities that others have simply overlooked: "Discovery consists of seeing what everybody has seen and thinking what nobody has thought."

Educational author George Kneller notes that such discoveries or reflective insights often appear to us at moments when we take time to reexamine our thinking to consider what circumstances or assumptions might be misconstrued.

Kneller observes:

> *Creativity, as has been said, consists largely of rearranging what we know in order to find out what we do not know. Hence, to think creatively, we must be able to look afresh at what we normally take for granted.*

American entrepreneur, business magnate, inventor and college drop-out Steve Jobs described creativity as "just connecting things." He explained:

> *When you ask creative people how they did something, they feel a little guilty because they didn't really do it. They just saw something. It seemed obvious to them after a while; that's because they were able to connect experiences they've had and synthesize new things.*

Former Harvard Business School professor and Harvard Business Review editor Theodore Levitt offers an important distinction between creativity and innovation, whereby: "Creativity is thinking up new things. Innovation is doing new things."

With regard to the former, American vocalist Judy Collins shares that:

> *I think people who are creative are the luckiest people on Earth. I know that there are no shortcuts, but you must keep your faith in something Greater than You, and keep doing what you love, and you will find the way to get it out of the world.*

W. Arthur ("Skip") Porter, a Texas business executive who served as Oklahoma Secretary of Science and Technology, characterized the innovation point as: "The pivotal point when talented and motivated people seek the opportunity to act on their ideas and dreams."

Celebrating Interests and Passions

Motivation expresses how much we really care about something...whether that is enough to cause us to actually do something about it. The evidence lies in our behaviors and actions regarding what we care most about, and the responsibilities we are prepared to recognize.

Our special interests and passions are precious assets that can lead us on wonderful voyages of exploration and opportunity. These expeditions of mind and action yield valuable lessons, skills and opportunities that enrich our lives and connect to who we are.

The human mind possesses a boundless capacity to care deeply about things beyond self...preferably, lots of somethings and someones. These passions guide and motivate us to experience and express our highest human potentials: to set goals; to meet challenges and seek excellence that set exceptional examples; to create music, art and literature that lift our intellect and spirits; and to believe in the power of worthwhile ideas and our abilities to make them real.

Passions arise from many sources. Sometimes they arise from exposures to a wide variety of people, ideas and experiences. On other occasions, they come to light through problems and obstacles we confront. They often relate to talents and other qualities we acknowledge in ourselves and others. Many are contagious and can be shared to forge bonds with kindred souls who share our values.

One big advantage of those who have passion about something is that they are often too preoccupied with what they care about to become sidetracked listening to experts who are all too willing to tell them otherwise. In fact, the very reason many people become successful is because they became interested in a possibility or discovery that nobody else paid much attention to or thought would work.

Pursuing passions stimulates our curiosity, inciting us to courses of inquiry and persistent action that reveal unexpected possibilities. In pursuit of new worlds of understanding, the late theoretical physicist and cosmologist Stephen Hawking observed: "Science is not only a discipline of reason, but also one of romance and passion."

Canadian jazz guitarist Randy Bachman appears to have a strong sense of what this pursuit of fulfillment means to him:

> *When one knows at an early age that their gift, talent use direction is musical, one tends to focus on that and let nothing interfere or impede the forward motion toward the end of that rainbow. And after 50-something rears of rockin' out, you still realize there is no end to that distant rainbow until one's last sunset.*

Celebrity talk show hostess, actress and movie producer Oprah Winfrey attributes the driving energy force behind her achievements as passion: "[T]he power that comes from focusing on what excites you."

Sometimes survival is not nearly enough to wish for. Consider the experiences of a small mollusk known as the sea squirt, for example. After swimming around early in life and eventually finding a permanent place such as a barnacle it can attach to, its thinking mission is then fully accomplished. The little squirt then absorbs its own brain which it no longer needs for nutrients to be rebuilt into other organs.

Even if no one, including ourselves, expects us to change the world, we do so every day by just taking up space. So, while we're here, we might as well do something useful, interesting and stimulating that causes us to evolve...to make something of rich opportunities to become something more.

Opening Up to Inspiration

However incomplete and inadequate our attempts to define it, inspiration is something needed to fill an otherwise human void. It is something that guides our quest for understanding and practicing higher values. It is something that reveals forgotten beauty of nature and wisdom. It is something that provides examples of excellence we can aspire to and learn from, including generosity, courage, creativity, tenacity and true-life achievements. It is something that arouses our senses...something you feel when it touches you, sometimes prompting you to touch back.

Inspiration is all things we can imagine, and much, much more.

How do we recognize it?

Sometimes it arrives in our consciousness as a thunder clap of *WOW!* or as a silent unexpected tear we shed when it softly touches our hearts. Sometimes it appears in the form of provident dreams upon which to construct marvelous thought castles of promise to house realities much larger than ourselves. Sometimes it is a force transmitted through bonds and connections of love and friendship that empower us… and often humble us as well.

Sometimes we are its agents.

Without meaning to we inspire others through shared experiences and lessons. Sometimes inspiration transforms other times it instructs. Sometimes it enriches a moment…at others, it influences a lifetime.

Oftentimes we let it inspiration find and recognize us.

It's great to receive gifts of inspirational wisdom and thoughtfulness from others with much experience and insight to share. I will pass along a just a few that you may appreciate as I do. First, from Oscar Wilde:

> *Keep love in your heart. A life without it is like a sunless garden when the flowers are dead.*

From Helen Keller:

> *The best and most beautiful things in the world cannot be seen or even touched—they must be felt with the heart.*

Walt Whitman:

> *Keep your face always toward the sunshine—and shadows will fall behind you.*

Rabindranath Tagore:

> *Clouds come floating into my life, no longer to carry rain or usher storm, but to add color to my sunset sky.*

Henry David Thoreau:

It's not what you look at that matters, its what you see.

Jim Elliot:

"Wherever you are—be all there.

And most important of all, from Omar Khayyam:

Be happy for this moment. This moment is your life.

Charting Some Final Thoughts...So Far

Sharing these exuberantly and deliciously self-indulgent thoughts and experiences has brought back countless personal memories, reflections and lessons learned. First and foremost, the time and thought I have devoted to writing this book is a selfish gift to myself. Through the process I have granted myself license and priority to explore enormously broad and diverse aspects of purposeful value-seeking guidance that I can continue to apply to my own life going forward.

As with my previous book, *Thinking Whole,* the exploratory journey I have experienced throughout this project has been a wonderfully selfish mind-expanding adventure. It has led me to rediscover and appreciate marvelously wise, thoughtful and often humorous observations about basic life lessons, values and travails. While there are many of these that I would not wish to repeat, there are none that I regret. All delivered me through marvelous adventures to realize a life by design that serves my unbounded wishes.

Writing this book has challenged me to determine the most important life lessons that I can confidently present to you, so that I can share them with myself with the same confident assurances. Some key priorities which readily come to mind are summarized in the chart that follows.

My ongoing life design plan is to keep practicing them until I get better.

Respect your own values and strengths			
Allow personal interests to introduce you to great opportunities.	Don't accept boundary limits. Give yourself room to work and grow.	Pursue success rather than approval. It is easier and more fun to achieve.	Take sensible and worthwhile risks. Security is only a myth anyway.
Continue to learn and grow			
Look at the big picture. Keep that image in mind when you design for parts.	Observe and apply lessons from nature. They never go out of style.	Trust your experience and intuition above conflicting advice from experts.	Learn to be your own worst critic. Don't waste good time on bad ideas.
Recognize the importance of relationships			
Your choices of partners may be your most important decision in life.	Be a team player. Put common goals above your ego and self interest.	Share ideas and credit others generously. You just can't afford not to.	Always try to be honest and fair. Trust is more important than popularity.

CPSIA information can be obtained
at www.ICGtesting.com
Printed in the USA
BVHW042308030722
641183BV00001B/73